Techniques
of
Algebra

HERBERT A. HOLLISTER
Bowling Green State University

Techniques
of
Algebra

Harper & Row, Publishers
New York, Hagerstown, San Francisco, London

To my wife and children

Sponsoring Editor: Charlie Dresser
Project Editor: Brenda Goldberg
Designer: Emily Harste
Production Supervisor: Will C. Jomarrón
Compositor: Santype International Limited
Printer and Binder: Halliday Lithograph Corporation
Art Studio: Danmark & Michaels Inc.

**Techniques
of
Algebra**

Library of Congress Cataloging in Publication Data

Hollister, Herbert A
    Techniques of algebra.

    Includes index.
    1. Algebra. I. Title.
QA154.2.H64   512.9   76-49548
ISBN 0-06-042861-9

A preliminary edition of this title was published previously.

Contents

Preface   vii
Foreword to the Instructor   ix

1. The Language and the Objects   1
    1. The Use of Letters to Name Numbers   1
    2. The Language of Sets   4
    3. The Real Number System   11
    4. The Order Properties of R 23
    5. Classification of Numbers   30

2. Linear Equations and Inequalities   33
    1. Linear Equations in One Variable   33
    2. More Complex Linear Equations in One Variable   39
    3. Literal Equations and Formulas   43
    4. Systems of Equations in Two Variables   46
    5. Systems of Equations in Three or More Variables   54
    6. Solving Linear Inequalities   58
    7. Problem Solving   69

3. The Arithmetic of Algebraic Expressions   79
    1. The Arithmetic of Polynomials   80
    2. The Product of Two Binomials   89

3. Factoring Polynomials   91
4. Simplifying Algebraic Fractions   99
5. Multiplication and Division of Algebraic Fractions   104
6. Addition and Subtraction of Algebraic Fractions   108
*7. The Factor Theorem   113

4. Exponents, Roots, and Radicals   117
1. Integral Exponents   117
2. Square Roots   126
3. Other Roots and Radicals   134
4. Decimal Approximations of Numbers Expressed Using Radicals   139
5. Rational Exponents   145
6. Summary of the Definitions and Laws Developed in this Chapter   151

5. Polynomial Equations and Inequalities   153
1. Solving Polynomial Equations by Factoring   153
2. Completing the Square   159
3. The Quadratic Formula   168
4. Polynomial Inequalities   172
5. Solving Quadratic Inequalities   178
6. Applications   185

6. Graphs   193
1. The Coordinate System and the Concept of a Graph   193
2. Lines   199
3. Determining Equations of Lines   206
4. Linear Variation   212
5. The Distance Formula, Circles   215
6. Graphs and Systems of Equations   222
7. Functions   227
8. Applications of Functions   238
*9. The Arithmetic of Functions   248
*10. Ellipses   259
*11. Parabolas   265
*12. Hyperbolas   274

Appendix: Summary of the Arithmetic of Signed Numbers   287
Answers to Odd-Numbered Exercises   291
Index   307
Index of Symbols   310

Preface

This text has been designed for use in a one-term course for college students who need to master the basic skills and principles of algebra. No previous algebraic background has been assumed, but students are expected to possess the usual computational skills.

Algebra is treated as generalized arithmetic whose language and basic governing principles are developed in Chapter 1. All of the techniques treated in the rest of the text are explained using these principles in the belief that students remember "what to do" much better when they understand "why they can do it."

In Chapter 2 students are introduced to problem solving using linear equations and inequalities, and in Chapters 3 and 4 they develop the manipulative skills needed to deal with more sophisticated algebraic expressions. Problem solving using polynomial equations and inequalities is treated in Chapter 5, and the basics of graphing are covered in Chapter 6. Each concept and technique is illustrated by a great number of examples— over 400 step-by-step computations and graphs—with every newly introduced technique justified by the basic algebraic principles or by concepts previously developed.

I would like to express my appreciation to my typist, Ms. Connie Wetzel, for her outstanding work and dedication; to Don Burdsall of Marion

Technical College, for class testing the first draft of the manuscript; to Vijaya Manepalli, Bryan Damecki, Bill Shaughnessy, and Jim Williams, for assisting me in class testing the material; to Imogene Krift and Marilyn Titkemeier, for preparation of the preliminary edition; to Mary Chambers, Linda Shellenbarger, and Kae Lea Main, for their secretarial assistance; and to Bruce Johnson for his work in preparing the answers.

Herbert A. Hollister

Foreword to the Instructor

In preparing a basic algebra text for college students, we were faced with the problem of trying to satisfy the needs of several types of courses. Students taking such courses enter with highly varied backgrounds—some have studied algebra previously but have forgotten what they learned or were unsuccessful in their previous studies, while others have never seen algebra before. To accommodate all such students, the text assumes only a background in those topics commonly covered in eighth-grade mathematics. Because most current elementary mathematics programs do cover the arithmetic of signed numbers, we felt that this topic was best covered in an appendix. Practice with arithmetic is provided in many of the exercise sets throughout the text.

In Chapter 1, we offer an approach that will suit both those instructors who wish to follow a more formal approach and those who prefer an intuitive introduction to numbers. The basic properties of real numbers are listed in a sequence of axioms and theorems with each statement followed by numerical examples. The number line is used to illustrate the laws governing inequalities and absolute value. This chapter should be viewed as a foundation for the techniques developed later. The instructor who wishes to present mathematics as a logical system can do so just by further developing the proofs of some of the theorems.

Chapter 2 deals with linear equations and inequalities in one variable; systems of linear equations in two, three, or four variables; and problem solving using the techniques developed. The chapter provides the student with a basic understanding of solving equations, solving problems, and a certain amount of practice in arithmetic. Although many authors do not treat systems of equations until after studying graphs, we include them at this point for sound pedagogical reasons. First, we wanted to get to problem solving early in the book. Because a certain degree of frustration or boredom is likely to set in when studying manipulative skills, students are better motivated if they see some practical applications early in their study. Since they do not need to master complicated manipulative techniques in order to solve linear equations and systems, these methods and their use in problem solving should be available to them in the first few weeks of the course. The problems in the text are somewhat different from the usual algebra problems and present some interesting challenges for students. They have not been classified as to "type" because, for example, it does students very little good to memorize "how to do 'mixture' problems." They need practice in analyzing problems, expressing relationships in mathematical statements, and using basic techniques to solve the resulting equations and inequalities.

In Chapters 3 and 4 the basic algebraic skills needed in further mathematical studies are developed. Notation and terminology as well as the techniques under consideration are explained and justified. Throughout the text we have used as a guide the belief that students remember "how" better if they understand "why."

Chapter 5 treats polynomial equations and inequalities using the techniques developed in Chapters 3 and 4 and then applies these new skills to more problem solving. Chapter 6 introduces students to graphs and functions. This chapter includes linear variation and a discussion of graphical solutions of systems of equations. The latter part of the chapter provides an introduction to conic sections for those instructors who wish to include this topic in a basic course.

Most of the exercise sets provide more problems than most students need or have time to do. Answers have been provided for the odd-numbered ones so that instructors have the option of assigning problems for which the answers are unavailable.

H.A.H.

1
The Language and the Objects

Every subject taught or studied has its own special language. In economics we use the terms profit, revenue, inflation, and balance of trade; biology introduces the words mammal, vertibrate, and protozoa; the behavioral sciences use the terms ego, peer pressure, and IQ. In this chapter we introduce some of the basic language of algebra. This language is not complicated nor sophisticated. It is, rather, a type of "shorthand" that makes it easier to write mathematical statements. The objects of primary concern in this course are just numbers and the language and symbolism are developed so that we can better understand and deal with numerical statements.

1. The Use of Letters to Name Numbers

For some students, algebra seems drastically different from arithmetic because algebra uses "letters" instead of "numbers." This impression is created at an early age and is sometimes nearly impossible to dispel. If you have this notion, then before you can succeed, you need to change your thinking and to get a more accurate view of algebra. In this course, we will primarily be concerned with numbers just as you were in arithmetic. In arithmetic, however, when you make statements about numbers, they are usually specific statements about specific numbers. For example,

$$14 + 3 = 17$$
$$11 < 19$$
$$\frac{2}{3} + \frac{1}{5} = \frac{13}{15}$$
$$\frac{1}{8} + \frac{1}{2} \neq \frac{1}{7}$$
$$14{,}312 + 7{,}215 = 21{,}527,$$

and

$$1{,}435{,}712 - 821{,}601 = 614{,}111$$

are all statements about specific numbers.

In each of the above sentences, every number is represented by a numeral and *each* number is named *specifically*.

Let us borrow from grammar and consider each numeral to be a proper noun because it names a number. Since the same number can have several names, there is a difference between numbers and their names. For example, .5, 1/2, 2/4, 11/22, and 100/200 all name the same number.

What do we do when we wish to discuss numbers in general? Must we use phrases such as "a given number" or "some number?" The sentence:

"The sum of any number and its double is three times the number."

is certainly much more complicated than the arithmetic sentences we wrote earlier. The difficulty is that the statement we made is a *general* statement about numbers instead of a specific statement about specific numbers. This statement assures us that, in particular,

$$7 + (2 \cdot 7) = 3 \cdot 7$$
$$19 + (2 \cdot 19) = 3 \cdot 19$$
$$4213.76 + (2 \cdot 4213.76) = 3 \cdot 4213.76$$

In order to write the general statement more simply, we can use a letter to represent numbers and write:

If N is a number, then $N + (2 \cdot N) = 3 \cdot N$.

The letter N in this case is a symbol that represents numbers. What numbers does it represent? The statement says that it represents all numbers. We can view N as a *pronoun*. Just as the word "he" can be used to represent any man or boy, a letter can be used to represent any number. *If you will just view letters in algebra the same way you do pronouns in grammar*, many difficulties will be avoided. You treat pronouns like nouns, treat letters like numerals. When we use a letter, say x or y, to represent numbers, we call the letter a *variable*. Symbols, such as 5, $\sqrt{2}$, and π, that represent particular numbers are called *constants*.

The purpose of mathematical symbolism is to enable us to express ideas and statements more simply. Once we master the symbolism, we are able to grasp the ideas more easily. You have probably been doing this for many years by using formulas. In writing down formulas we frequently choose initial letters to help us remember the quantity being represented. The formula is shorter and easier to remember than the verbal statements.

Example 1

Area of a rectangle = length × width

$A = L \times W$ □

Example 2

Area of a triangle = $\frac{1}{2}$ × base × height

$A = \frac{1}{2} \times B \times H$ □

Example 3

Circumference of a circle = π × diameter

$C = \pi \times D$ □

The last example includes a special number that is not easily represented using the numerals from elementary arithmetic. The Greek letter "π" pronounced "pi" is universally used to represent the constant ratio between the circumference of a circle and its diameter; that is, if C is the circumference of a circle and D its diameter, then $\pi = C/D$. By using advanced numerical techniques we can show that π is close to 3.1416. *It is not equal to 3.1416* but 3.1416 is a sufficiently close approximation for many applications and practical computations.

The symbol π is a proper noun or numeral just like 5, 19, and 304.62 because it names a specific number. When this symbol arises in algebra, treat it just like you do 7 or 34 or any other numeral.

If you remember that using letters to represent numbers just provides us with a quicker, shorter, simpler way of making statements and you treat the letters just as you would treat any other names for numbers, you will avoid most symbolic difficulties.

EXERCISES

Write formulas for the following and list the quantity represented by each letter. You may have to look up some of them.

1. The perimeter of a rectangle.
2. The area of a parallelogram.
3. The area of a trapezoid.
4. The volume of a rectangular solid.
5. The area of a circle.
6. The surface area of a cylinder.
7. The volume of a cylinder.
8. The surface area of a right circular cone.
9. The volume of a cone.
10. The volume of a pyramid.

2. The Language of Sets

With the introduction of "modern math" into the elementary and secondary schools, some students were told that they were learning "set theory." This was not quite true. Formal set theory is a rigorous, highly abstract part of mathematics, too sophisticated for study at the elementary or secondary level. Many current programs have, however, benefited from including some study of sets but some rather formal and rigorous programs have confused students with a great deal of unnecessary abstraction.

The *language* of sets is very useful in mathematics in that it enables us to condense and simplify many mathematical statements and hence make them easier to grasp. In this section we will develop enough of this language to help us simplify and clarify some of the concepts that will be developed later in the text. We will not attempt to develop set theory but will be content to discuss certain basic concepts and establish some notational conventions.

Intuitively, the word "set" means the same as the words "collection," "class," "aggregate," and "bunch." The "things" in a set are called its *elements*. Thus, for example, *Treasure Island* is an element of the set of all books written by Robert Louis Stevenson, and Bowling Green State University is an element of the set of all universities in Ohio.

Just as we use letters to name numbers, we can use letters to name sets and elements of sets. Usually we use lower case letters to indicate elements, and capital letters to denote sets.

If A is a set and p is an element of A, we can abbreviate the sentence "p is an element of A" by "$p \in A$." The symbol "$\in$" is read "is an element of," "belongs to," or "is in."

Example 1

If E denotes the set of even numbers, then $6 \in E$. ☐

If A and B are sets such that every element of A is an element of B, we say that A is a *subset* of B. A is a *proper* subset of B if A is a subset of B but A is not equal to B.

The sentence "A is a subset of B" can be abbreviated by "$A \subseteq B$." The symbol "$\subseteq$" is read "is a subset of" or "is contained in." Do not confuse the symbols "$\subseteq$" and "$\in$;" they have different meanings.

Example 2

If N denotes the set of all numbers and E denotes the set of even numbers, then $E \subseteq N$. In fact, E is a proper subset of N.

Note that E is *not* an element of N; the elements of N are numbers whereas E is a *set* of numbers. ☐

To negate these symbols "$\in$" and "$\subseteq$" we follow the customary practice of drawing a line through the symbol. That is, $p \notin A$ means "p is not an element of A," and $A \nsubseteq B$ means "A is not a subset of B."

Once the elements of a set are determined, the set is completely determined. Thus one of the simplest methods of describing a set is just to list its elements, and we usually do this within a set of braces.

Example 3

The set of vowels is {a, e, i, o, u}. ☐

Sets can also be described by "set-builder notation" using a set of braces, a letter, and a sentence containing that letter. The set so described is the set of objects that when substituted for that letter yield a true sentence.

Example 4

{x: x is a professional hockey player}

is read

"The set of all x such that x is a professional hockey player." ☐

In the above example we would expect all possible replacements for x to be people. In some cases the possible replacements for x may not be clear from the context, and we specify them by requiring in our notation that x be an element of a specific set.

Example 5

Let B denote the set of all Bowling Green State University students. Then

$\{x \in B: x \text{ has a 3.20 average}\}$

is the set of all students at Bowling Green State University who have a 3.20 average. $\square$

Using set-builder notation or some other descriptive method, we might describe a set that has no elements.

Example 6

Let $\mathcal{N}$ denote the set of all numbers. Then

$\{x \in \mathcal{N}: x \neq x\}$

is a set with no elements because every number is equal to itself. $\square$

Example 7

Similarly,

$\{y \in \mathcal{N}: y + 1 = y\}$

is a set with no elements because adding 1 to every number changes it. $\square$

Two sets are the same if they have the same elements. The two sets in these examples are the same and are equal to every set with no elements. The set with no elements is called the *empty set* or the *null set* and is denoted by $\varnothing$.

If A and B are sets, then the *union of A and B* is denoted by $A \cup B$ and is defined by

$A \cup B = \{x: x \in A \text{ or } x \in B\}$

The union of more than two sets is defined analogously.

The *intersection* of two sets, A and B, is denoted by $A \cap B$ and is defined by

$A \cap B = \{x: x \in A \text{ and } x \in B\}$

The intersection of more than two sets is defined analogously.

Example 8

Suppose

$A = \{1, 2, 4, 5, 7\} \quad \text{and} \quad B = \{1, 4, 7, 9, 10\}$

Then

$$A \cup B = \{1, 2, 4, 5, 7, 9, 10\}$$

and

$$A \cap B = \{1, 4, 7\} \ \square$$

If $A \cap B = \emptyset$, then we say that A and B are *disjoint*.

If A and B are sets, then the *difference* of A and B is denoted by $A - B$, and is defined by

$$A - B = \{x : x \in A \text{ and } x \notin B\}$$

Example 9

Suppose M denotes the set of all men in the world, W denotes the set of all women in the world, and K denotes the set of all married people in the world. Then

$$M \cap W = \emptyset$$

that is, M and W are disjoint. Note also that:

$M \cap K$ is the set of married men.
$M \cap W$ is the set of married women.
$M - K$ is the set of single men.
$W - K$ is the set of single women. $\square$

At this point a remark is in order regarding the use of parentheses. Parentheses are used in mathematics, as in grammar, for punctuation. When we enclose an expression in parentheses, we mean to consider the quantity inside the parentheses as one object rather than by its components, and other symbols are related to such quantities as they would be to single objects.

Example 10

By $A \cup (B - C)$ we mean the union of the set A with the set $B - C$. $\square$

Example 11

By $5 \times (7 + 4)$ we mean the product of 5 with the sum of 7 and 4. $\square$

Example 12

By $(8 + 10) \div 3$ we mean the quotient obtained by dividing the sum of 8 and 10 by 3; that is, $18 \div 3$. $\square$

The concept of *equality is essentially a logical concept.* Two objects are equal if and only if *they are identical,* that is, they are the same object. Whenever we write $A = B$, we mean that A and B are two names for the same object. This narrow view of equality will prevail throughout the text.

If we know that $A = B$, then we can substitute B for A in any expression involving A and the result will have the same meaning because B *means the same* as A. It is this principle that enables us to "add the same number to both sides of an equation" or to substitute $3 + 2$ for 5 in any numerical expression. We can use this substitution principle with sets, numbers, or any objects we desire.

Example 13

If A, B, and C are sets such that $A = B$, then $A \cup C = B \cup C$ and $A \cap C = B \cap C$. $\square$

Example 14

If $A = B$ and $A \subseteq C$, then $B \subseteq C$. $\square$

Example 15

If $A = B$ and A and C are disjoint, then B and C are disjoint. $\square$

Example 16

If $3x - 4 = x$, then $(3x - 4) + 4 = x + 4$. $\square$

Example 17

If $4x = 12$, then $4x/4 = 12/4$. $\square$

There are a few basic principles that are very useful in dealing with sets. We list them below without proof but will illustrate some of them by examples and drawings.

(a) If A and B are sets such that $A \subseteq B$ and $B \subseteq A$, then $A = B$.

Statement (a) is one of the most commonly used tools from set theory. To show that two sets are the same, the easiest approach in many cases is to show that each is a subset of the other. In a formal development of set theory, this statement is one of the fundamental axioms, called the *Axiom of Extension.* The assertions below can be proved using this statement and the definitions.

(b) If A, B, and C are sets such that $A \subseteq B$ and $B \subseteq C$, then $A \subseteq C$.

(c) If A is a set, then $\varnothing \subseteq A$.

The only way that $\varnothing$ could fail to be a subset of A is if $\varnothing$ contains some element not in A. Since $\varnothing$ contains no elements, this is impossible. Thus $\varnothing \subseteq A$.

(d) If A and B are sets, then $A \cup B = B \cup A$ and $A \cap B = B \cap A$.

Example 18

Let

$$A = \{1, 2, 3, 4, 5\} \quad \text{and} \quad B = \{3, 4, 5, 6, 7\}$$

Then

$$A \cup B = \{1, 2, 3, 4, 5, 6, 7\} = B \cup A$$

and

$$A \cap B = \{3, 4, 5\} = B \cap A \quad \square$$

Statement (d) is true primarily because the order of listing elements in a set does not matter.

(e) If A, B, and C are sets, then

$$A \cup (B \cup C) = (A \cup B) \cup C$$
$$A \cap (B \cap C) = (A \cap B) \cap C$$
$$A \cap (B \cup C) = (A \cap B) \cup (A \cap C)$$
$$A \cup (B \cap C) = (A \cup B) \cap (A \cup C)$$

In order to better illustrate statements (b) and (e), it is convenient to introduce the idea of a *Venn diagram*.

These diagrams frequently are used to describe the relationships among sets both intuitively and geometrically. The usual procedure is to first draw a rectangle to represent some set containing all elements under discussion and we call this the *universal set* or *universe of discourse*. We then draw all sets under consideration as sets of points inside circles or other closed curves inside the rectangle.

Example 19

The union and intersection of sets A and B are represented by drawings in Figures 1-1 and 1-2. The union is shaded in the first drawing, and the intersection is shaded in the second. $\square$

Figure 1–1 $A \cup B$

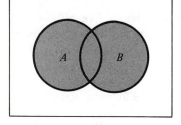

Figure 1–2 $A \cap B$

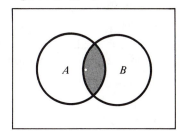

Figure 1-3 $A \cap B = \phi$

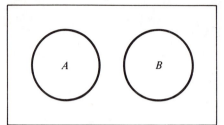

Figure 1-4 $A \subseteq B$

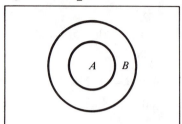

Figure 1-5 $A \cup (B \cap C)$

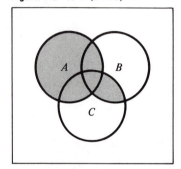

Figure 1-6 $(A \cup B) \cap (A \cup C)$

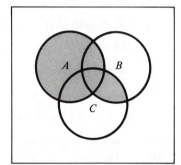

Example 20

The drawing in Figure 1-3 indicates that A and B are disjoint, and Figure 1-4 indicates that $A \subseteq B$. ☐

We can use Venn diagrams to help our intuition but we must realize that drawings do not constitute proofs.

Example 21

By shading $A \cup (B \cap C)$ as in Figure 1-5 and shading $(A \cup B) \cap (A \cup C)$ as in Figure 1-6, we would expect that these sets were the same. We could not, however, claim to have a proof that such an equation would always hold. ☐

Since this is not a rigorous course in set theory, we will accept justifications made by means of Venn diagrams, providing the diagrams represent the general situation.

EXERCISES

1. Let $A = \{a, b, c, d, e, f\}$ and $B = \{a, c, e, g, h, k\}$. List the following sets.
   (a) $A \cup B$ (b) $A \cap B$ (c) $A - B$
   (d) $B - A$ (e) $A - (A \cap B)$ (f) $(A - B) \cup (B - A)$

2. Let A and B be arbitrary sets. Draw Venn diagrams to represent the following.
   (a) $A - B$ (b) $B - A$ (c) $A - (A \cap B)$ (d) $(A - B) \cup (B - A)$

3. Let A and B be arbitrary sets. Draw Venn diagrams to represent the following.
   (a) $(A \cup B) - C$ (b) $A \cup (B - C)$
   (c) $A \cap (B - C)$ (d) $A \cap (B \cap C)$
   (e) $(A \cap B) \cap C$ (f) $A \cap (B \cup C)$
   (g) $A \cup (B \cap C)$ (h) $(A - B) \cup (B - C) \cup (C - A) \cup (A \cap B \cap C)$

4. Let A, B, and C be arbitrary sets. Use Venn diagrams to illustrate and justify the following statements.
   (a) $A \cup (B \cup C) = (A \cup B) \cup C$ (b) $A \cap (B \cap C) = (A \cap B) \cap C$
   (c) $(A \cap B) \subseteq A$ (d) $A \subseteq (A \cup B)$
   (e) $A \cap (B \cup C) = (A \cap B) \cup (A \cap C)$

5. Let A, B, C, and D be sets. Decide whether the following are true or false. You may wish to use Venn diagrams to help make your decision.
   (a) $A \cap (A \cup B) = A$
   (b) $A \cup (A \cap B) = A$
   (c) If $A \subseteq B$ and $B \cap C = \emptyset$, then $A \cap C = \emptyset$.
   (d) If $A \subseteq B$ and $A \subseteq C$, then $A \subseteq B \cap C$.
   (e) If $A \cup B \subseteq A \cap B$, then $A = B$.
   (f) $(A - B) \cap C = (A \cap C) - B$
   (g) If $A \cup B = A \cup C$, then $B = C$.
   (h) If $A \subseteq B$ and $C \subseteq D$, then $A \cap C \subseteq B \cap D$.
   (i) $(A - B) \cap B = \emptyset$
   (j) $A - (B \cap C) = (A - B) \cap (A - C)$
   (k) $A - (B \cap C) = (A - B) \cup (A - C)$
   (l) $A - (B \cup C) = (A - B) \cap (A - C)$

3. The Real Number System

As we remarked in the introduction, this course is primarily concerned with numbers. We are, however, more concerned with general properties and behavior of numbers than we are with specific numbers. The set of numbers we will use is generally known as the *real number system*. There is a larger system, known as the complex numbers, which are of use in higher mathematics, physics, and engineering, but they will not concern us in this course. In this text the word "number" means "real number." We will not attempt to give a rigorous development of the real number system but will be content to provide the student with an intuitive, geometric, arithmetic grasp of the system and its properties. We will, however, organize the basic properties of the system into a set of axioms and a sequence of theorems. This will enable you to refer to the properties more easily and to see how one property may depend on others. A few proofs have been included to provide illustrations of how the theorems can be derived from the basic properties, the axioms. Proofs, after all, are just explanations of why a statement is true. Throughout this text we will be providing explanations of *why* certain statements are true and *why* certain methods can be used because we believe that you will *remember best* if you *understand why* the fact that you are studying is true. The explanations will have, as their foundation, the assertions listed in this section.

In writing numerical expressions we adopt the following standard notational conventions.

I. Multiplication will usually be indicated either by a dot "·" or by juxtaposition, that is, by placing the numbers side by side.

Example 1

By $5 \cdot 3$ we mean the product of 5 and 3. The expression $(5)(3)$ also means their product. □

Example 2

If we represent two numbers by letters x and y, we usually write xy for the product of x and y. □

II. Multiplication and division take precedence over addition and subtraction.

Example 3

$5 \cdot 3 + 4$ means to add the product of 5 and 3 to the number 4, that is, first multiply 5 times 3 and then add 4. If we wished to indicate the product of 5 with the sum of 3 and 4, we would use parentheses and write $5 \cdot (3 + 4)$. □

Example 4

$pq + mn$ means $(p \cdot q) + (m \cdot n)$. □

Example 5

$x + y/z$ means $x + (y/z)$. We do the division before the addition. □

III. Multiplication and division are to be performed in the order in which they occur as are addition and subtraction.

Example 6

$7 \cdot 8 \div 2$ means $(7 \cdot 8) \div 2 = 56 \div 2 = 28$. □

Example 7

6 ÷ 3 ÷ 2 means (6 ÷ 3) ÷ 2 = 2 ÷ 2 = 1. □

Example 8

14 − 5 + 3 means (14 − 5) + 3 = 9 + 3 = 12. □

Example 9

17 − 6 − 4 means (17 − 6) − 4 = 11 − 4 = 7. □

When you are writing expressions, make sure that you indicate clearly what you mean. If parentheses would clarify your meaning, use them. Notational conventions save us some writing but should not lead to confusion.

Intuitively, the real numbers may be thought of as the points of a line, as in Figure 1-7.

Figure 1-7

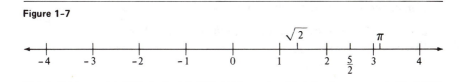

Numbers may also be considered as directed distances from the origin or "zero" point. Intuitively, a number p is less than a number q if p is "to the left" of q on the line.

Besides comparing numbers, we expect to be able to do arithmetic, that is, to add, subtract, multiply, and divide numbers, and these processes, along with the ordering of numbers, are described and governed by certain fundamental principles. These principles are presented in the form of axioms, lemmas, and theorems. The axioms are the fundamental assumptions that we make about the number system. Essentially we are saying "If the number system is to behave as our intuition and experience expect it to behave, then the following basic statements must hold." Once we choose the axioms, then the lemmas and theorems all can be proved by proceeding logically from the axioms. We will not provide proofs of most of these statements since we are primarily concerned with your ability to *use* these properties rather than your ability to derive or prove them. You should make sure that you understand the principles under discussion.

Axioms for the Real Numbers

The real numbers, denoted by R, is a set with two operations, addition and multiplication, denoted by + and ·, respectively, which satisfies the following axioms.

AXIOM 0

If a and b are real numbers, then $a + b$ and $a \cdot b$ are real numbers. (Closure Axiom)

AXIOM 1

If a and b are real numbers, then

$$a + b = b + a \quad \text{and} \quad a \cdot b = b \cdot a \qquad \text{(Commutative Laws)}$$

Example 10

$$15 + 24 = 24 + 15 \quad \text{and} \quad (15)(24) = (24)(15) \quad \square$$

AXIOM 2

If a, b, and c are real numbers, then

$$a + (b + c) = (a + b) + c \quad \text{and} \quad a \cdot (b \cdot c) = (a \cdot b) \cdot c$$
(Associative Laws)

Example 11

$$16 + (13 + 20) = (16 + 13) + 20 \qquad \text{because } 16 + 33 = 29 + 20$$
$$5 \cdot (7 \cdot 3) = (5 \cdot 7) \cdot 3 \qquad \text{because } 5 \cdot 21 = 35 \cdot 3 \quad \square$$

AXIOM 3

If a, b, and c are real numbers, then

$$a \cdot (b + c) = a \cdot b + a \cdot c \quad \text{and} \quad (a + b) \cdot c = a \cdot c + b \cdot c$$
(Distributive Laws)

Example 12

$$5(20 + 7) = 5 \cdot 20 + 5 \cdot 7 \qquad \text{since } 5(27) = 100 + 35$$

In fact, it is this principle that lets us multiply $5 \cdot 27$ by multiplying $5 \cdot 20$ and $5 \cdot 7$ and then adding. $\square$

AXIOM 4

There is a unique real number 0 such that if a is any real number, then

$$a + 0 = 0 + a = a \quad \text{and} \quad a \cdot 0 = 0 \cdot a = 0$$

AXIOM 5

There is a unique real number 1, different from 0, with the property that if a is any real number, then

$$a \cdot 1 = 1 \cdot a = a$$

AXIOM 6

If a is a real number, then there is a unique real number, denoted by $-a$ and called the *negative of a*, with the property that

$$a + (-a) = (-a) + a = 0$$

Example 13

The negative of 15 is -15 because $15 + (-15) = 0$. The negative of -12 is 12 because $(-12) + (12) = 0$. $\square$

AXIOM 7

If a and b are nonzero real numbers, then

$$a \cdot b \neq 0$$

AXIOM 8

There is a relation called "less than" denoted by " $<$ " defined on R which satisfies the following conditions:

(i) For every pair of real numbers a and b, exactly one of the following holds: $a < b$, $a = b$, or $b < a$.
(ii) If a, b, and c are real numbers such that $a < b$ and $b < c$, then $a < c$.

The relation "less than" is used to define another relation "greater than" by saying that a is greater than b if and only if b is less than a. We denote "greater than" by " $>$."

If a and b are real numbers we say that $a \leq b$ if $a < b$ or $a = b$. The symbol " $\leq$ " is read "less than or equal to." The meaning of the symbol " $\geq$ " is defined in a similar manner.

A number is said to be *positive* if it is greater than 0 and *negative* if it is less than 0.

AXIOM 9

If a, b, and c are real numbers such that $a < b$, then $a + c < b + c$.

Example 14

Since $4 < 11$, $4 + 5 < 11 + 5$ and $4 + (-6) < 11 + (-6)$. $\square$

Example 15

Since $-4 < 2$, $(-4) + 3 < 2 + 3$; that is, $-1 < 5$. $\square$

(See the Appendix for a review of the arithmetic of signed numbers.)

AXIOM 10

If a, b, and c are real numbers such that $a < b$ and $0 < c$, then $ac < bc$.

Example 16

Since $5 < 7$ and $0 < 3$, $5 \cdot 3 < 7 \cdot 3$, that is, $15 < 21$. $\square$

Example 17

Since $-6 < -2$ and $0 < 5$, $(-6)5 < (-2)5$; that is, $-30 < -10$.
(The fact that $(-6)5 = -30$ and $(-2)5 = -10$ will be justified by Theorem 3.) $\square$

If a and b are real numbers, then a/b is the unique real number whose product with b is a if such a unique real number exists. The number a/b is called a *divided by* b, or the *quotient* of a and b. We also use the usual divisional sign " $\div$." Remember that "$a \div b$," $\dfrac{a}{b}$, and "a/b" mean the same thing.

Example 18

$\left(\dfrac{72}{9}\right)9 = 72$ $\square$

Example 19

$\dfrac{266}{19} = 14$ because $14 \cdot 19 = 266$ $\square$

Example 20

$\left(\dfrac{143}{\pi}\right)\pi = 143$ $\square$

AXIOM 11

If a and b are real numbers and $b \neq 0$, then a/b exists in R; that is, there is a unique real number whose product with b is a.

AXIOM 12

If S is a nonempty subset of R and a is a real number such that every element of S is less than a, then there is a smallest real number b which is greater than or equal to every element of S.

Axiom 12 is called the "least upper bound property." An upper bound of a set is just a number greater than or equal to every number in the set and Axiom 12 asserts that any set having an upper bound in R must have a smallest or least upper bound in R. It is this property that assures us that every point on the line is associated with a real number and conversely, that every real number is associated with a point. Furthermore, this property completely characterizes the real number system and distinguishes it from the rational number system and others. We will not use Axiom 12 in this text but we include it for the sake of completeness.

Using the axioms, we can prove the following lemmas and theorems. We will not give proofs of most of them but will do so for the first three because many students have never seen a justification of the "Law of Signs." We suggest that the student attempt to provide proofs for some of the other assertions or at least parts of them. We will illustrate all of the assertions by examples.

LEMMA 1

If a is a real number, then

$$(-1)a = -a$$

Proof:

$$
\begin{aligned}
a + (-1) \cdot a &= 1 \cdot a + (-1) \cdot a && \text{because } a = 1 \cdot a \text{ by Axiom 5} \\
&= [1 + (-1)] \cdot a && \text{by the distributive law, Axiom 3} \\
&= 0 \cdot a && \text{because } 1 + (-1) = 0 \text{ by Axiom 6} \\
&= 0 && \text{by Axiom 4}
\end{aligned}
$$

Thus

$$a + (-1) \cdot a = 0$$

Since $-a$ is the only number whose sum with a is 0, $(-1) \cdot a$ must be $-a$, and, therefore,

$$(-1) \cdot a = -a$$

Example 21

$(-1)(7) = -7$ □

Example 22

$(-1)\pi = -\pi$ □

Do you recall the first time you were told that $(-3) \cdot (-5) = 15$? For many students, it does not seem possible that the product of two negative numbers could be positive. Some high school courses justify this by several intuitive and physical examples, but such arguments are quite artificial. That $(-3) \cdot (-5)$ must be 15 is a direct consequence of the first six axioms for R. Thus if we are to have a number system satisfying these basic properties, we must have $(-3)(-5) = 15$.

Before proving the general case of the Law of Signs in Theorem 3, we will prove a special case as a lemma which will then enable us to easily prove the more general case.

LEMMA 2

$(-1) \cdot (-1) = 1$

Proof:

$$(-1) \cdot (-1) = -(-1)$$
$$= 1$$

by Lemma 1 where (-1) is a
The negative of $(-a)$ is just a
for any number a because of
the uniqueness required in
Axiom 4.

THEOREM 3 (The Law of Signs)

If a and b are real numbers, then

$$-(a \cdot b) = (-a) \cdot b = a \cdot (-b) \quad \text{and} \quad (-a) \cdot (-b) = a \cdot b$$

Proof:

$$-(a \cdot b) = (-1) \cdot (a \cdot b)$$
$$= [(-1) \cdot a] \cdot b$$
$$= (-a) \cdot b$$

by Lemma 1
by associativity, Axiom 2
by Lemma 1

Similarly,

$$(-a) \cdot b = [(-1) \cdot a] \cdot b$$
$$= [a \cdot (-1)] \cdot b$$
$$= a \cdot [(-1) \cdot b]$$
$$= a \cdot (-b)$$

by Lemma 1
by commutativity, Axiom 1
by associativity, Axiom 2
by Lemma 1

Therefore

$$-(a \cdot b) = (-a) \cdot b = a \cdot (-b)$$

and the first part of the theorem is proved. The second part follows by applying Lemmas 1 and 2:

$$
\begin{aligned}
(-a) \cdot (-b) &= [(-1) \cdot a] \cdot [(-1) \cdot b] & \text{by Lemma 1} \\
&= [a \cdot (-1)] \cdot [(-1) \cdot b] & \text{by commutativity, Axiom 1} \\
&= a \cdot [(-1) \cdot \{(-1) \cdot b\}] & \text{by associativity, Axiom 2} \\
&= a \cdot ([(-1)(-1)] \cdot b) & \text{by associativity, Axiom 2} \\
&= a \cdot (1 \cdot b) & \text{by Lemma 2} \\
&= a \cdot b & \text{by Axiom 5}
\end{aligned}
$$

Therefore

$$(-a) \cdot (-b) = a \cdot b$$

Example 23

$$(3)(-4) = -(3 \cdot 4) = -12 \ \square$$

Example 24

$$(-5)(2) = -(5 \cdot 2) = -10 \ \square$$

Example 25

$$(-4)(-6) = (4)(6) = 24 \ \square$$

In basic arithmetic, subtraction is defined in terms of addition.

Example 26

$$47 - 19 = 28 \qquad \text{because } 28 + 19 = 47 \ \square$$

In general, $a - b$ is the number whose sum with b is a, that is,

$$(a - b) + b = a$$

Example 27

$$21 - 15 = 6 \qquad \text{because } 6 + 15 = 21 \ \square$$

Example 28

$$19 - 26 = -7 \qquad \text{because } (-7) + (26) = 19 \ \square$$

THEOREM 4

If a and b are numbers, then

$$a - b = a + (-b)$$

Proof:

$$[a + (-b)] + b = a + [(-b) + b] = a + 0 = a \quad \square$$

PROPOSITION 5

If a, b, and c are real numbers, then

$$a \cdot (b - c) = ab - ac \quad \text{and} \quad (a - b) \cdot c = ac - bc$$

Proof: Exercise

From Axiom 11, we know that division of a real number by a nonzero real number is always defined in R; that is, if x and y are real numbers and $y \neq 0$, then x/y is a real number. Recall that, by definition, $(x/y)y = x$.

Is division by 0 possible? Consider, for example, the possibility of dividing 1 by 0. Is there a real number whose product with 0 is 1? Since the product of 0 with any number is 0, no such number exists. In general, if p is a nonzero real number, then there is *no* number whose product with 0 is p, and therefore, division of p by 0 is impossible. Is it possible to divide 0 by 0? Since $0 \cdot 0 = 0$ and $1 \cdot 0 = 0$, there is *more than one* real number whose product with 0 is 0. In order for 0/0 to have meaning, 0/0 would have to be the *unique* number whose product with 0 is 0, and there is no such unique number.

In the above discussion we have actually proved the following assertion:

PROPOSITION 6

If r is a real number, then $r/0$ has no meaning; that is, there is no unique real number whose product with 0 is r.

Our reason for choosing the fraction line for the division symbol becomes clear as we develop the properties of division. All the usual techniques for computing with fractions hold just because a fraction is, in fact, a quotient.

Frequently when a student first encounters fractions in arithmetic, he gets the idea that the arithmetic of fractions is different in some way from the arithmetic of whole numbers. That is, he expects to have different rules for operating with fractions than for operating with other numbers. Such an assumption or conclusion is a handicap in arithmetic. Fractions are, after all, just a way of representing numbers. Numbers represented by fractions have to behave according to the same laws as other numbers, that is, according to the principles governing the real numbers. Therefore, the basic properties of the arithmetic of numbers represented by fractions are the same as the basic properties of arithmetic for all numbers. The computational techniques developed are, in fact, just techniques for computing with numbers expressed in fractional form, that is, techniques for using numerators and denominators of fractions in the computation of sums, differences, products, and quotients of numbers expressed as fractions. Since the fraction line means division, all these techniques are derivable and provable using just the meaning of division and the basic properties of the arithmetic of the real number system. We have put these together in the following theorem so that the student may see the relationship between these techniques and division. We will discuss these principles in more detail when we study algebraic fractions.

THEOREM 7

Let a, b, c, and d be real numbers with $b \neq 0$ and $d \neq 0$. The following all hold.

(i) $(a/b)b = a$

(ii) $a/1 = a$

(iii) $b/b = 1$

(iv) $a(1/b) = a/b$

(v) $(a/b)c = ac/b$

(vi) $(a/b)/d = a/bd$

(vii) $a/b = ad/bd$

(viii) $a/b = c/d$ if and only if $ad = bc$.

(ix) $(a/b)(c/d) = ac/bd$

(x) $(a/b) \div (c/d) = (a/b)(d/c)$ if $c \neq 0$.

(xi) $-(a/b) = (-a)/b = a/(-b)$

(xii) $a/b + c/b = (a + c)/b$

(xiii) $a/b - c/b = (a - c)/b$

(xiv) $a/b + c/d = (ad + bc)/bd$

(xv) $a/b - c/d = (ad - bc)/bd$

Proof:

We will not provide proofs of all of the above statements but will give justifications for some of them and leave the rest for the student. All of the principles are illustrated by Example 29.

Principle (i) follows from the definition of division in that a/b is defined to be the number whose product with b is a.

Principle (iv) can be established just by showing that $a(1/b)$ is the number whose product with b is a. If we form the product, we see that this is, in fact, true.

$$\left[a\left(\frac{1}{b}\right)\right]b = a\left[\left(\frac{1}{b}\right)b\right] = a(1) = a$$

In order to establish principle (xii) we need to show that $a/b + c/b$ is the number whose product with b is $a + c$. Most of the other principles in this theorem can be established by similar arguments. Some of them require a bit more reasoning using the previously established principles. The student should attempt some of them.

Example 29

The following numerical statements illustrate the corresponding parts of Theorem 7.

(i) $(4/3)3 = 4$; $(35/5)5 = 35$

(ii) $-26/1 = -26$; $(4/5)/1 = 4/5$

(iii) $9/9 = 1$; $(2/3)/(2/3) = 1$

(iv) $6(1/5) = 6/5$; $(-7)(1/9) = -7/9$

(v) $(4/5)3 = 12/5$; $(-6/7)(4) = -24/7$

(vi) $(2/3)/5 = 2/15$; $(4/7)/3 = 4/21$

(vii) $2/3 = (2 \cdot 7)/(3 \cdot 7) = 14/21$; $9/4 = (9 \cdot 2)/(4 \cdot 2) = 18/8$

(viii) $2/3 = 8/12$ because $2 \cdot 12 = 3 \cdot 8$

$4/7 \neq 5/8$ because $4 \cdot 8 \neq 7 \cdot 5$

(ix) $(2/3)(4/5) = 8/15$; $(3/4)(2/3) = 6/12$

(x) $(2/3) \div (5/7) = (2/3)(7/5) = 14/15$

(xi) $-(4/5) = (-4)/5 = 4/(-5)$

(xii) $4/3 + 7/3 = (4 + 7)/3 = 11/3$

(xiii) $8/5 - 11/5 = (8 - 11)/5 = (-3)/5$

(xiv) $\dfrac{2}{3} + \dfrac{4}{5} = \dfrac{2 \cdot 5 + 3 \cdot 4}{3 \cdot 5} = \dfrac{10 + 12}{15} = \dfrac{22}{15}$

(xv) $\dfrac{5}{8} - \dfrac{2}{3} = \dfrac{5 \cdot 3 - 8 \cdot 2}{8 \cdot 3} = \dfrac{15 - 16}{24} = \dfrac{(-1)}{24}$ □

EXERCISES

1. Use the definition of division to decide whether the following are true or false.

(a) $224/7 = 32$ (b) $323/17 = 19$

(c) $629/27 = 27$ (d) $280/15 = 19$

(e) $115/5 = 23$ (f) $962/26 = 27$

2. Illustrate each part of Theorem 7 by letting $a = 14$, $b = 11$, $c = 7$, and $d = 5$.

3. Illustrate each part of Theorem 7 by letting $a = -18$, $b = 6$, $c = -2$, and $d = -3$.
4. Justify part (vii) of Theorem 7 by showing that $(a/b)(bd) = ad$. [*Hint:* $(a/b)(bd) = ([(a/b)b]d)$
5. Justify part (x) of Theorem 7 by showing that

$$[(a/b)(d/c)](c/d) = a/b$$

6. Justify part (xii) of Theorem 7 by showing that

$$(a/b + c/b)b = a + c$$

7. Justify part (xiv) of Theorem 7 by using parts (vii) and (xii). [*Hint:* $a/b = ad/bd$]
*8. Complete the proof of Theorem 7.

4. The Order Properties of *R*

Along with being able to add, subtract, multiply, and divide real numbers, we can also compare them, that is, determine when one number is less than another. This comparability is of primary importance in many applications. We may wish to compare bowling scores, test scores, prices of automobiles, costs of manufacturing swimsuits, or temperatures. The principles governing ordering and inequalities are given by Axioms 8, 9, 10, and 12, and in the consequences of these axioms.

Intuitively, Axiom 9 assures us that if we add the same number to both sides of an inequality, we obtain an inequality in the same direction or sense. Axiom 10 asserts that if we multiply both sides of an inequality by a *positive* number, we obtain an inequality in the same direction. The following theorem contains the other basic principles governing inequalities and these can be proved using the axioms. As before, we will not provide proofs but will illustrate each part of the theorem by examples and illustrations.

THEOREM 8

Let a, b, c, and d be real numbers. Then the following hold.

   (i) If $a < b$, then $-b < -a$.
  (ii) If $0 < a$ and $0 < b$, then $0 < ab$.
 (iii) If $a < 0$ and $b < 0$, then $ab > 0$.
  (iv) If $a < b$ and $0 < c$, then $a/c < b/c$.
   (v) If $a < b$ and $c < 0$, then $ac > bc$ and $a/c > b/c$.
  (vi) If $a \neq 0$, then $0 < a \cdot a$.
 (vii) $0 < 1$
(viii) $c > 0$ if and only if $1/c > 0$.
  (ix) Let $bd > 0$. Then $a/b < c/d$ if and only if $ad < bc$.

Example 1

The following numerical statements illustrate the corresponding parts of Theorem 8.

(i) Since $2 < 5$, $-5 < -2$.
  Since $-3 < 7$, $-7 < 3$.
  Since $-8 < -2$, $2 < 8$.
(ii) Since $0 < 7.2$ and $0 < 3$, $0 < 21.6$.
(iii) Since $2 < 5$ and $0 < 7$, $2/7 < 5/7$.
  Since $-8 < -4$ and $0 < 2$, $-8/2 < -4/2$, that is, $-4 < -2$.
(iv) Since $3 < 5$ and $-2 < 0$, $3/(-2) > 5/(-2)$, that is, $-(3/2) > -(5/2)$.
  Since $-10 < -4$ and $-8 < 0$, $-10/(-8) > -4/(-8)$, that is, $5/4 > 1/2$.
(v) Since $-6 < 0$ and $-2/3 < 0$, $(-6)(-2/3) > 0$, that is, $4 > 0$.
(vi) Since $5 \neq 0$, $0 < 25$.
  Since $-3 \neq 0$, $0 < 9$.
(viii) Since $5 > 0$, $1/5 > 0$.
  Since $1/19 > 0$, $19 > 0$.
  Since $4/7 > 0$, $1/(4/7) > 0$, that is, $7/4 > 0$.
(ix) Since $2 \cdot 3 > 0$ and $7 \cdot 2 < 3 \cdot 9$, $7/3 < 9/2$.

To decide whether or not $42/13 < 51/15$, since $13 \cdot 15 > 0$, we compute $42 \cdot 15$ and $13 \cdot 51$ and compare. Since $42 \cdot 15 = 630 < 663 = 13 \cdot 51$ we know $42/13 < 51/15$. □

Proof of Theorem 8:

We will not prove each part of the theorem but will justify some of them. The student should attempt to prove the others. Principle (i) is established by using Axiom 9 twice. Just add $-a$ to both sides and then add $-b$ to both sides.

To prove (ii), notice that if

$$\frac{a}{c} \not< \frac{b}{c}$$

then either

$$\frac{a}{c} = \frac{b}{c} \quad \text{or} \quad \frac{b}{c} < \frac{a}{c}$$

Then, since $0 < c$, either

$$\left(\frac{a}{c}\right)c = \left(\frac{b}{c}\right)c \quad \text{and} \quad a = b$$

or

$$\left(\frac{b}{c}\right)c < \left(\frac{a}{c}\right)c \quad \text{and} \quad b < a$$

Since both of these are impossible, it must be that

$$\frac{a}{c} < \frac{b}{c} \quad \square$$

Number lines can also be used to illustrate parts of Theorem 8.

Example 2

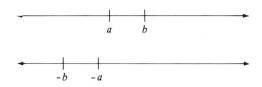

(i) If *a* is to the left of *b*, then −*b* is to the left of −*a*.

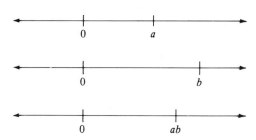

(ii) If *a* and *b* are both positive, so is *ab*.

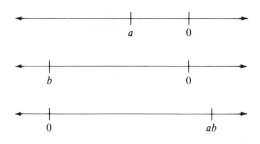

(iii) If both *a* and *b* are negative, then *ab* is positive.

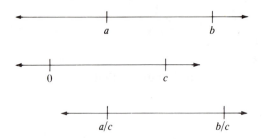

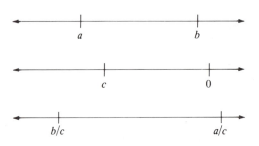

(iv) If *a* is to the left of *b* and *c* is positive, then *a/c* is to the left of *b/c*.

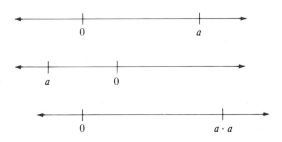

(v) If *a* is to the left of *b* and *c* is negative, then *b/c* is to the left of *a/c*.

(vi) If $a \neq 0$, then $a \cdot a$ is positive.

Axiom 10 and principles (ii) and (iv) from Theorem 8 will be used a great deal in Chapter 2 to solve algebraic inequalities. They tell us that multiplying or dividing both sides of an inequality by a *positive* number *preserves* the direction of the inequality, whereas multiplying or dividing both sides of an inequality by a *negative* number *reverses* the direction of the inequality.

The concept of absolute value corresponds with the intuitive idea of the *size* of a number. The absolute value of a number can also be viewed as the distance of the number from 0 on the number line.

Example 3

The absolute value of 7 is 7 because the distance from 0 to 7 is 7 units.

The absolute value of −5 is 5 because the distance from 0 to −5 is 5 units.

The concept is made precise by the following definition:

If *a* is a real number, the *absolute value* of *a* is denoted by $|a|$ and defined by

$$|a| = a \quad \text{if} \quad a \geq 0$$
$$|a| = -a \quad \text{if} \quad a < 0 \quad \square$$

Example 4

Since $\pi > 0$, $|\pi| = \pi$.
Since $-5 < 0$, $|-5| = -(-5) = 5$.
Since $4\pi - 13 < 0$, $|4\pi - 13| = -(4\pi - 13) = 13 - 4\pi$. $\square$

Note that $|a| \geq 0$ for every number *a*. If a number is positive, then it is equal to its absolute value; if it is less than 0, then its absolute value is its negative. (See Figure 1-8.)

Figure 1-8

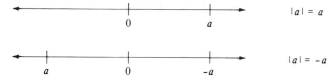

Example 5

Determine

$$|14/9 - 5/3|$$

We need to know whether this number is positive or negative.

$$14/9 - 5/3 = 14/9 - 15/9 = -1/9$$

Thus

$$|14/9 - 5/3| = |-1/9| = 1/9 \quad \square$$

The following theorem contains the basic principles governing the concept of absolute value.

THEOREM 9

Let p and q be real numbers. Then the following hold.

(i) $|p| = |-p|$
(ii) $-|p| \leq p \leq |p|$
(iii) $|pq| = |p| \cdot |q|$
(iv) If $p \geq 0$, then $|q| \leq p$ if and only if $-p \leq q \leq p$.
(v) $|p + q| \leq |p| + |q|$
(vi) $|p - q| \leq |p| + |q|$

The theorem can be proved by using the definition and properties previously established; for example, (v) can be proved using (iv) and (ii). We will not provide a proof of Theorem 9 but will, rather, illustrate portions by means of accompanying number lines.

(i) $|4| = |-4|$ since both are 4 units from 0.

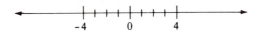

(iii)

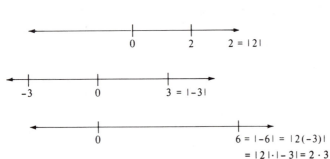

(iv) $-7 \leq q \leq 7$ if and only if $0 \leq |q| \leq 7$.

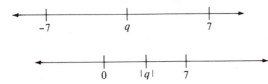

(v) $|2 + 3| = |5| = 5 = 2 + 3 = |2| + |3|$

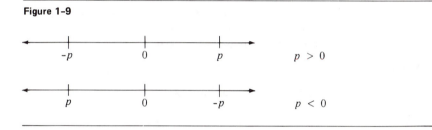

$|(-2) + 3| = |1| = 1 \leq |-2| + |3| = 2 + 3 = 5$

$|(-5) + 2| = |-3| = 3 \leq |-5| + |2| = 5 + 2 = 7$

$5 = |-5| = |(-3) + (-2)| = |-3| + |-2| = 3 + 2$

Figures 1-9 and 1-10 provide general illustrations of principles (i) and (iv). We leave further illustrations as exercises.

Figure 1-9

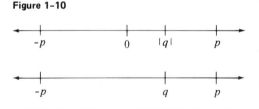

Figure 1-10

EXERCISES

1. Use Theorem 8 and the axioms to decide if the following are true or false.

(a) $16/32 < 94/188$
(b) $51/33 > 85/55$
(c) $72/19 < 145/39$
(d) $4.23/7.12 > 531/1124$
(e) $-312/54 < -727/124$
(f) $14/(-112) > 42/(-326)$
(g) $96/(-113) < 72/(-101)$
(h) $-47/31 > 59/47$
(i) $16 + \pi < 32 + \pi$
(j) $-4 + \pi > -7 + \pi$
(k) $\pi < 2\pi$
(l) $4739/18 < 4786/18$
(m) $1/27 < 1/25$
(n) $5/37 > 5/39$

2. Use Axioms 8 and 9 to explain why the sum of two positive numbers is positive and why the sum of two negative numbers is negative.
3. Explain why the product of a positive number and a negative number is negative.
4. Decide whether the following are true or false.
   (a) $|3 - \pi| = 3 - \pi$
   (b) $|7 - (183/476)| = 7 - (183/476)$
   (c) $|43 - 19| = |19 - 43|$
   (d) $|\pi/.00001| = \pi/.00001$
   (e) $|7 - 9| = |7| - |9|$
   (f) $|7\pi - 22| = 7\pi - 22$
*5. Use Theorem 8 and explain why $|a/b| = |a|/|b|$ for any numbers a and b (provided that $b \neq 0$).
*6. If $|a| < |b|$ and $a \neq 0$, what can be said about $|1/a|$ and $|1/b|$?
*7. Complete the proof of Theorem 8.
*8. Complete the proof of Theorem 9.

5. Classification of Numbers

It is convenient to classify numbers in the following traditional manner and to assign standard symbols as names for these sets of numbers.

The *natural numbers*, denoted by N, is another name for those numbers you may have called positive whole numbers; that is,

$$N = \{1, 2, 3, 4, \ldots\}$$

The set N can be axiomatized by the following assertions:

   (i) $N \subseteq R$
   (ii) $1 \in N$
   (iii) If $a \in N$ and $b \in N$, then $(a + b) \in N$ and $(a \cdot b) \in N$.
   (iv) If K is any subset of R satisfying (i), (ii), and (iii), then $N \subseteq K$.

This is another way of saying that N is the smallest set of numbers containing 1 that is closed under addition and multiplication. (Closed means that whenever two numbers are in the set, their sum and product are also.)

The *integers*, denoted by Z, is another name for what you may have called whole numbers; that is,

$$Z = \{\ldots, -2, -1, 0, 1, 2, 3, \ldots\}$$

The *rational numbers*, denoted by Q, are those real numbers which can be expressed as quotients (ratios) of integers:

$$Q = \{a/b : a \in Z, b \in Z, \text{ and } b \neq 0\}$$

Notice that $N \subseteq Z \subseteq Q \subseteq R$.

Real numbers that are not rational are said to be *irrational*; the number π is an example of such a number. In Chapter 4 we will discuss many other irrational numbers.

An integer p is said to be *prime* if $p > 1$ and p is not evenly divisible by any positive integers other than itself and 1; that is, if a and b are positive integers such that $ab = p$, then $a = p$ and $b = 1$ or $b = p$ and $a = 1$.

Example 1

The number 5 is prime because the only positive integers dividing 5 evenly are 1 and 5. $\square$

Using our definitions of N and Z, we can, with *quite a lot of work*, prove the following *fundamental theorem of arithmetic*.

THEOREM 10

Every integer greater than 1 is either prime or can be expressed uniquely as the product of prime integers.

Example 2

If we write $70 = 2 \cdot 5 \cdot 7$, then we have expressed 70 as a product of primes. Theorem 10 assures us that this can be done in only one way (except, of course, for the order of the primes). $\square$

Example 3

$504 = 2 \cdot 2 \cdot 2 \cdot 3 \cdot 3 \cdot 7$ $\square$

Example 4

$462 = 2 \cdot 3 \cdot 7 \cdot 11$ $\square$

A rational number p/q is said to be expressed in *lowest terms* if

(i) q is positive
(ii) p and q are not both evenly divisible by a common prime.

By repeated use of Theorem 7, part (vii) $[a/b = ad/bd]$ and the fundamental theorem of arithmetic, it follows that any rational number may be expressed in lowest terms and the expression will be unique.

Example 5

The fraction 4/6 is not in lowest terms because 4 and 6 are both divisible by 2. By part (vii) of Theorem 7 $[a/b = ad/bd]$ we know

$$\frac{4}{6} = \frac{2 \cdot 2}{3 \cdot 2} = \frac{2}{3} \quad \square$$

Example 6

The fraction 14/27 is in lowest terms because the only primes dividing 14 are 2 and 7 and neither of these divides 27. □

Notice that in reducing fractions to lowest terms, we need only look for primes that divide the numerator and denominator.

Example 7

$$\frac{105}{231} = \frac{35 \cdot 3}{77 \cdot 3} = \frac{35}{77} = \frac{5 \cdot 7}{11 \cdot 7} = \frac{5}{11} \quad \square$$

EXERCISES

1. Is Q closed under addition and multiplication? Justify your answer.
2. List the primes between 1 and 100.
3. Decide whether or not 1021 is prime.
4. Use Theorem 10 to decide whether or not $7 \cdot 17 \cdot 19 \cdot 13 = 23 \cdot 31 \cdot 71$.
5. Reduce the following to lowest terms.

   (a) 15/35   (b) 24/90

   (c) 121/77   (d) 16/100

   (e) 49/27   (f) 76/57

   (g) 85/119   (h) 130/169
*6. What is the largest divisor you would have to try to decide whether or not 211 is prime? Justify your answer.
*7. Can you generalize your result from Exercise 6?

Linear Equations and Inequalities

To many people, algebra means "solving" equations. Such an impression is not quite accurate but it is true that developing techniques for solving equations and inequalities is one of the most important objectives of a course in algebra. In this chapter we deal with the simplest equations and inequalities, those that are called *linear*. We will see why the word linear is used when we study graphs in Chapter 6.

The student should realize that mathematics does not exist just to solve mathematician's puzzles. Historically, mathematics has its roots in the natural sciences and practical applications. Complex problems from the natural sciences, the behavioral sciences, and the fields of administration and finance can be solved only by using advanced mathematical techniques.

Students who are taking this course in order to prepare themselves for the problems that arise in their fields of specialization should be especially interested in the applications discussed in the last section of the chapter.

1. Linear Equations in One Variable

What is an equation? An equation is a simple sentence using the linking verb "equals." In the sentence

"The sum of three and two equals five."

the subject is "The sum of three and two," the verb is "equals," and the predicate noun is "five." Mathematical symbols enable us to write this sentence in a shorter form:

"$3 + 2 = 5$"

As you might expect, most of the equations we write will be written symbolically. When we are dealing with applications, many of these equations will arise from verbal statements about the problem situation. In such situations, our first task will be to formulate the symbolic equations.

Recall from Chapter 1 that there are two basic types of symbols that represent numbers. We have *numerals* such as 2, 5, 1342, 47/93, and π that name particular numbers and that can be viewed as proper nouns. The second type of symbolic representation for numbers is by *variables* such as a, b, x, y, z, p, q, that can be viewed as pronouns. For example, when we write

"$3 + x = 5$"

the symbol "x" is a pronoun representing a number. In this case, our knowledge of arithmetic assures us that 2 is the only number that x can represent and still make the sentence true. If we write

$$a + b = b + a$$

we have a sentence that holds for all possible numbers that we could substitute for a and b. An equation that holds for all possible meaningful substitutions is called an *identity*. An equation that holds for only part of the possible meaningful substitutions is called a *conditional equation*. In our study of the laws of arithmetic, we made several statements that were identities. In this chapter and those that follow, we will be concerned primarily with conditional equations.

To solve a conditional equation containing one variable, say x, means to determine all numbers which when substituted for the variable yield a true statement. These numbers are called *solutions*.

Example 1

The equation

$$(-4) + x = 19$$

is a conditional equation. If we substitute 23 for x in this equation, the resulting sentence is true. All other substitutions yield false statements. Thus 23 is the only solution to the equation. $\square$

Example 2

The equation

$$8x = 12$$

is a conditional equation. If we substitute $\frac{3}{2}$ for x, the resulting sentence is true. All other substitutions yield false statements. Thus $\frac{3}{2}$ is the only solution to the equation. □

Example 3

The equation

$$3x = \frac{15}{2}$$

is a conditional equation. The only solution is $\frac{5}{2}$. □

A major portion of this book is devoted to methods of finding solutions of equations. One fundamental principle underlies all of these techniques. Loosely speaking it says: "If we do the same thing to both sides of a true equation, the resulting statement is also true."

Example 4

$$3 + 2 = 5$$

is known to be true. Thus

$$(3 + 2)7 = 5 \cdot 7$$

is also true as are

$$(3 + 2) - 8 = 5 - 8$$

and

$$\frac{3 + 2}{9} = \frac{5}{9} \quad □$$

In this book, we take a narrow view of equality. Equality means "sameness" and when we write

"$3 + 2 = 5$"

we mean that "$3 + 2$" names the same number that "5" does. *If we have two names for the same number, we can interchange them in any algebraic expression and obtain two expressions that represent the same number.* This is all that underlies the "do the same thing to both sides" idea.

Example 5

Solve the equation

$$(-5) + y = 31$$

We would like to end up with an equation of the form $y =$ _____ where the blank is filled in with the number which, when substituted for y, yields a true statement. We could employ a form such as

$$(-5) + \underline{\hspace{1cm}} = 31$$

and then try to decide what number to use in filling the blank. This is a common practice in elementary schools to help students learn the facts of arithmetic. By clever guessing we might decide that the blank should be filled with 36 and we would be correct. If we use the first form

$$(-5) + y = 31$$

then by "adding 5 to both sides" we obtain a new equation:

$$5 + [(-5) + y] = 5 + 31$$

The associative law enables us to rewrite this as

$$[5 + (-5)] + y = 5 + 31$$

Now, via arithmetic, we obtain

$$0 + y = 36$$

and finally

$$y = 36$$

If we substitute 36 for y in the original equation we have

$$(-5) + 36 = 31$$

which is true. □

Example 6

In a similar fashion, suppose we have

$$5y = 34$$

Again we would like to obtain an equation of the form $y =$ _____. By analyzing this equation we see that if we "multiply both sides" by $\frac{1}{5}$ we have

$$\tfrac{1}{5}(5y) = \tfrac{1}{5}(34)$$

Then

$$(\tfrac{1}{5} \cdot 5)y = \tfrac{34}{5}$$
$$(1)y = \tfrac{34}{5}$$
$$y = \tfrac{34}{5} \quad \text{or} \quad y = 6\tfrac{4}{5} \quad \text{or} \quad y = 6.8$$

Substituting in the original equation we have

$$5(\tfrac{34}{5}) = 34$$

which is true.

In this example, we also could have "divided both sides" by 5 and obtained the equation

$$\frac{5y}{5} = \frac{34}{5}$$

which would yield

$$y = \frac{34}{5} \ \square$$

Similarly, in Example 5, we could have subtracted (-5) from both sides to obtain the desired solution.

Our techniques provide us with all possible solutions but we must "test" them to see if they actually work. In the examples given thus far, checking by substitution is very simple because the equations are simple. As we progress to more complicated equations, the checking computations become more involved. *In order to be sure that your solution is valid, you must try it.*

Example 7

Solve the equation $x + .3 = .9$ and give reasons for steps.

$x + .3 = .9$	Given
$(x + .3) + (-.3) = .9 + (-.3)$	Adding $-.3$ to both sides
$x + [.3 + (-.3)] = .6$	Associative law and arithmetic
$x + 0 = .6$	Arithmetic
$x = .6$	Arithmetic

Checking:

$$.6 + .3 = .9 \ \square$$

Example 8

$y + 7 = 3.2$	Given
$(y + 7) - 7 = 3.2 - 7$	Subtracting 7 from both sides.
$y = -3.8$	Arithmetic

Checking:

$$(-3.8) + 7 = 3.2 \quad \square$$

Example 9

$$\frac{2}{3}z = 41 \qquad \text{Given}$$
$$(\tfrac{3}{2})(\tfrac{2}{3}z) = (\tfrac{3}{2})41 \qquad \text{Multiplying both sides by } \tfrac{3}{2}.$$
$$[(\tfrac{3}{2})(\tfrac{2}{3})]z = \tfrac{123}{2} \qquad \text{Associativity and arithmetic}$$
$$1z = \tfrac{123}{2} \qquad \text{Arithmetic}$$
$$z = \tfrac{123}{2} \quad \text{or} \quad z = 61.5$$

Checking:

$$\frac{2}{3} \cdot \frac{123}{2} = \frac{246}{6} = 41 \quad \square$$

Example 10

$$4x = 1.9 \qquad \text{Given}$$

$$\frac{4x}{4} = \frac{1.9}{4} \qquad \text{Dividing both sides by 4.}$$

$$x = .475 \qquad \text{Arithmetic}$$

Checking:

$$4(4.75) = 1.9 \quad \square$$

In the above examples we have provided justification for each step. Although it is not necessary to write such justifications in order to solve an equation, the student should be able to justify each step to himself. Before proceeding to the exercises, the reader should study the following examples and provide a reason for each step.

Example 11

$$x + .5 = 1.2$$
$$(x + .5) + (-.5) = 1.2 + (-.5)$$
$$x + [.5 + (-.5)] = .7$$
$$x + 0 = .7$$
$$x = .7$$

Checking:

$$.7 + .5 = 1.2 \quad \square$$

Example 12

$$x - 3 = \tfrac{4}{5}$$
$$(x - 3) + 3 = \tfrac{4}{5} + 3$$
$$x = 3\tfrac{4}{5}$$

Checking:

$$3\tfrac{4}{5} - 3 = \tfrac{4}{5} \ \square$$

EXERCISES

1. Solve the following equations. Write out justifications for each step. Check your solutions to be sure they are correct.
   - (a) $x + .7 = 9.3$
   - (b) $y - 92 = 11$
   - (c) $1.5 = z + .04$
   - (d) $w + \tfrac{3}{4} = \tfrac{2}{3}$
   - (e) $x - \tfrac{4}{5} = \tfrac{11}{9}$
   - (f) $5y = 95$
   - (g) $\tfrac{1}{2}z = 95$
   - (h) $\tfrac{2}{3}x = 16$
   - (i) $4y = -18$
   - (j) $-6z = 138$
2. Solve the following equations.
   - (a) $x - \tfrac{2}{17} = \tfrac{1}{4}$
   - (b) $y + \tfrac{5}{11} = \tfrac{1}{8}$
   - (c) $z + .01 = \tfrac{1}{16}$
   - (d) $.5x = \tfrac{2}{3}$
   - (e) $\tfrac{4}{7}y = \tfrac{9}{10}$
   - (f) $.004x = 11$
   - (g) $\tfrac{2}{5}y = \tfrac{17}{19}$
   - (h) $\tfrac{11}{5}z = .02$
   - (i) $.06 + x = -1.47$
   - (j) $y - 11.98 = -1.274$
3. Solve the following equations.
   - (a) $.08x = 1.93$
   - (b) $\tfrac{4}{3}y = \tfrac{5}{7}$
   - (c) $x - \tfrac{7}{3} = -\tfrac{11}{15}$
   - (d) $y + \tfrac{9}{11} = \tfrac{5}{14}$
   - (e) $.03x = 9.7$
   - (f) $94y = .003$
   - (g) $\tfrac{21}{17} + y = \tfrac{94}{13}$
   - (h) $\tfrac{21}{17}x = \tfrac{94}{13}$
   - (i) $\tfrac{31}{7}x = \tfrac{14}{51}$
   - (j) $\tfrac{31}{7} + y = \tfrac{14}{51}$

2. More Complex Linear Equations in One Variable

The "do the same thing to both sides" technique can be applied several times in solving an equation.

Example 1

$3x + 2 = 9$	Given
$(3x + 2) - 2 = 9 - 2$	Subtracting 2 from both sides.
$3x = 7$	Arithmetic
$\dfrac{3x}{3} = \dfrac{7}{3}$	Dividing both sides by 3.
$x = \dfrac{7}{3}$	Arithmetic

Checking:

$$3\left(\frac{7}{3}\right) + 2 = 7 + 2 = 9 \;\square$$

Example 2

$5x = 2x + 8$	Given
$(-2x) + (5x) = (-2x) + [2x + 8]$	Adding $(-2x)$ to both sides.
$[(-2) + 5]x = [(-2x) + 2x] + 8$	Associative and distributive laws
$3x = 8$	Arithmetic
$\dfrac{3x}{3} = \dfrac{8}{3}$	Dividing both sides by 3.
$x = \dfrac{8}{3}$	Arithmetic

Checking:

$$5\left(\frac{8}{3}\right) = 2\left(\frac{8}{3}\right) + 8$$

is equivalent to

$$\frac{40}{3} = \frac{16}{3} + 8$$

which is true. $\square$

Example 3

$13x + 2 = 4x + 14$	Given
$(-4x) + [13x + 2] = (-4x) + [4x + 14]$	Adding $-4x$ to both sides.
$[(-4x) + 13x] + 2 = [(-4x) + 4x] + 14$	Associativity
$9x + 2 = 0 + 14$	Arithmetic
$9x + 2 = 14$	Arithmetic
$(9x + 2) - 2 = 14 - 2$	Subtracting 2 from both sides.
$9x = 12$	Arithmetic
$\dfrac{9x}{9} = \dfrac{12}{9}$	Dividing both sides by 9.
$x = \dfrac{4}{3}$	Arithmetic

Checking:

$$13\left(\frac{4}{3}\right) + 2 = \frac{52}{3} + 2 = \frac{58}{3} \quad \text{and} \quad 4\left(\frac{4}{3}\right) + 14 = \frac{16}{3} + 14 = \frac{58}{3} \;\square$$

The reader should provide the justification for each step in the following examples. At times two steps are combined into one.

Example 4

$$
\begin{array}{ll}
3x + 2 = 5x - 9 & \text{Given} \\
(-5x) + (3x + 2) = (-5x) + (5x - 9) & \\
-2x + 2 = -9 & \\
-2x = (-9) + (-2) & \\
-2x = -11 & \\
2x = 11 & \\
x = \dfrac{11}{2} &
\end{array}
$$

Checking:

$$3\left(\frac{11}{2}\right) + 2 = \frac{33}{2} + 2 = \frac{37}{2}$$

$$5\left(\frac{11}{2}\right) - 9 = \frac{55}{2} - 9 = \frac{37}{2} \;\square$$

Example 5

$$
\begin{array}{ll}
4x + 13 = 2x + 91 & \text{Given} \\
(-2x) + (4x + 13) = 91 & \\
2x + 13 = 91 & \\
2x = 78 & \\
x = 39 &
\end{array}
$$

Checking:

$$4(39) + 13 = 156 + 13 = 169$$
$$2(39) + 91 = 78 + 91 = 169 \;\square$$

Example 6

$$
\begin{array}{ll}
\frac{2}{3}x + \frac{5}{4} = \frac{1}{7}x - \frac{4}{5} & \text{Given} \\
(-\frac{1}{7}x) + \frac{2}{3}x = (-\frac{4}{5}) + (-\frac{5}{4}) & \\
(-\frac{1}{7} + \frac{2}{3})x = -(\frac{4}{5} + \frac{5}{4}) & \\
(-\frac{1}{7} + \frac{2}{3})x = -\left[\dfrac{16 + 25}{20}\right] &
\end{array}
$$

$$\left[\frac{(-3)+14}{21}\right]x = -\frac{41}{20}$$

$$\frac{11}{21}x = -\frac{41}{20}$$

$$x = \frac{21}{11}\left[-\frac{41}{20}\right]$$

$$x = -\frac{861}{220}$$

Checking:

$$\frac{2}{3}\left(-\frac{861}{220}\right) + \frac{5}{4} = -\frac{287}{110} + \frac{5}{4} = -\frac{574}{220} + \frac{275}{220} = -\frac{299}{220}$$

$$\frac{1}{7}\left(-\frac{861}{220}\right) - \frac{4}{5} = -\frac{123}{220} - \frac{4}{5} = -\frac{123}{220} - \frac{176}{220} = -\frac{299}{220} \ \square$$

The method followed in Examples 4–6 is typical of the methods you should eventually be able to use. In several cases, two or more steps were combined into one. If you cannot follow this procedure then fill in the single steps. After you become more accustomed to working with equations, such techniques will come naturally to you. Do not strive for them. *There is no great virtue in brevity.* Write down each step that you cannot do easily in your head.

EXERCISES

1. Solve the following equations. Write down each step and check your solutions.
   (a) $3x + 4 = 10$
   (b) $2x - 5 = 7$
   (c) $4y + 9 = 11$
   (d) $4z - 7 = 2z + 9$
   (e) $4x - 7 = 25$
   (f) $6x + 15 = 3x - 27$
   (g) $4x + 5 = 6x + 10$
   (h) $17x + 31 = 25x - 91$
   (i) $14x + 18 = 6x + 20$
   (j) $241x + 92 = 71x + 7$
2. Solve the following equations. Be sure to check your solutions.
   (a) $.2x + 3 = .5x - 6$
   (b) $7y - .5 = 2y + 4$
   (c) $8x - 7 = 5x + \frac{2}{3}$
   (d) $\frac{4}{5}x + \frac{7}{15} = \frac{2}{3}x - \frac{5}{9}$
   (e) $4x + .31 = 9x - .72$
   (f) $\frac{5}{6}x - \frac{3}{10} = \frac{7}{12}x + \frac{11}{30}$
3. Solve the following equations. Check your solutions.
   (a) $1.6x + 9.2 = .42x - 7.93$
   (b) $\frac{3}{7}x + \frac{7}{3} = \frac{7}{3}x + \frac{3}{7}$
   (c) $y/3 + 2y/5 = 7/10$
   (d) $\frac{4}{5}x - \frac{2}{3} = \frac{9}{7}x + \frac{5}{9}$
   (e) $31.46x + 92.3 = 7.41x - 15$
   (f) $\frac{4}{7}x + \frac{2}{3} = \frac{7}{10}x - \frac{5}{4}$
4. Solve the following equations.
   (a) $\frac{2}{3}x - \frac{7}{2} = \frac{5}{6}x + \frac{1}{3}$
   (b) $.7x + 3.2 = .4x + 3.8$
   (c) $72.3x - 19.7 = 35.6x - 11.4$
   (d) $\frac{5}{19}x + \frac{3}{19} = \frac{6}{19}x - \frac{5}{19}$
   (e) $320x + 421 = 280x - 39$
   (f) $41x + 300 = 62x + 120$
   (g) $3042x - 773 = 242x + 427$
   (h) $.0032x - .76 = .0016x - .12$
5. Does the equation $x = x + 2$ have any solutions? If not, why not?
6. Write three equations that do not have solutions.

3. Literal Equations and Formulas

Frequently we encounter equations, such as formulas, that include several quantities, all or some of which are represented by letters. Such equations are sometimes called *literal* equations.

Example 1

$A = lw$	the formula for the area of a rectangle
$p = 2l + 2w$	the formula for the perimeter of a rectangle
$d = rt$	the relationship between distance, rate, and time
$i = Prt$	the formula for computing interest when P denotes principle, r denotes rate, and t denotes time
$C = 2\pi r$	the formula for the circumference of a circle where r denotes the radius and π is a special constant (that is sometimes approximated by 3.1416 or by 22/7) $\square$

It is sometimes quite useful to solve such equations for one quantity in terms of the others.

Example 2

The formula $A = lw$ expresses the area in terms of the length and width. We may, however, wish to obtain a formula that expresses length in terms of area and width. To do so we can just solve the equation $lw = A$ for l by treating l as a variable and A and w as constants. In this case, if we divide both sides by w we have

$$\frac{lw}{w} = \frac{A}{w} \quad \text{and} \quad l = \frac{A}{w} \ \square$$

Example 3

Suppose we wish to express the length of a rectangle in terms of the width and perimeter. We know that

$$2l + 2w = p$$

Subtracting $2w$ from both sides yields

$$2l = p - 2w$$

Dividing both sides by 2 yields

$$l = \frac{p - 2w}{2} \quad \text{or} \quad l = \frac{p}{2} - w \ \square$$

Example 4

To express the radius of a circle in terms of the circumference, we solve $C = 2\pi r$ for r:

$$2\pi r = C$$

$$\frac{2\pi r}{2\pi} = \frac{C}{2\pi}$$

$$r = \frac{C}{2\pi} \,\square$$

In solving literal equations we use exactly the same techniques as in solving numerical equations. We just treat one of the letters as a variable and the others as we would any number.

Sometimes we have an equation involving two variables and it is useful to solve for one in terms of the other.

Example 5

Solve the equation $4x + 8y = 5$ for x in terms of y.

$$4x + 8y = 5$$

$$4x = 5 - 8y$$

$$\frac{4x}{4} = \frac{5 - 8y}{4}$$

$$x = \frac{5}{4} - 2y \,\square$$

Example 6

Solve the equation $5x + 7y = 8$ for y in terms of x.

$$5x + 7y = 8$$

$$7y = 8 - 5x$$

$$\frac{7y}{7} = \frac{8 - 5x}{7}$$

$$y = \frac{8}{7} - \frac{5}{7}x \,\square$$

EXERCISES

1. (a) Use the formula $d = rt$ to determine how far an automobile travels if it averages 55 mph for 4.3 hr.
   (b) Solve the formula for r in terms of d and t.

(c) If a runner runs 440 yd in 56 sec, what is his average rate?

(d) If an airplane travels 2430 mi in 5.4 hr, what is the average rate to the nearest mile per hour?

(e) Solve the formula for t in terms of d and r.

(f) How long does it take an automobile to go .01 mi at 55 mph? Express your answer to the nearest second. (It may be of interest that .01 mi is 52.8 ft.)

(g) How long does it take an airplane to go 5 mi at 500 mph? Again, express your answer to the nearest second.

(h) Spacecraft carrying astronauts to and from the moon reenter our atmosphere at 25,000 mph. How long does it take them to travel 10 mi? Express your answer to the nearest second.

2. (a) Determine the interest on $1200 invested at 8 percent per year for 7 months.

(b) What is the interest on $3200 at 9 percent per year for $5\frac{1}{2}$ months?

(c) Solve the formula $i = Prt$ for P, for r, and for t.

(d) If you are receiving $48 every 3 months on savings receiving 5 percent per year, how much have you invested in your savings account?

(e) If you are receiving $84 every 6 months on $3000, what is your annual rate of interest?

(f) If you are paying $30 interest every month on a loan of $5000, what is your annual rate of interest?

3. (a) Determine the perimeter of a rectangle of length 4.7 in. and width 2.6 in.

(b) What is the width of a rectangle with perimeter 35 cm and length 11 cm?

(c) What is the width of a rectangle with length 4.2 in. and area 10.5 sq in.?

4. The area of a triangle is given by the formula $A = \frac{1}{2}bh$.

(a) What is the area of a triangle with base 7.2 cm and height 3.1 cm?

(b) Solve the formula for b in terms of A and h. Solve it for h in terms of b and A.

(c) What is the base of a triangle with height 5 in. and area 24 sq in.?

(d) What is the height of a triangle with area 9.2 sq cm and base 11.5 cm?

5. The area of a trapezoid is given by the formula $A = \frac{1}{2}(b_1 + b_2)h$ where h denotes the height and b_1 and b_2 are the bases.

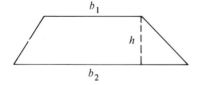

(a) Determine the area of a trapezoid with height 7.2 cm having bases of length 8.4 cm and 6.6 cm.

(b) Solve the formula for h, b_1, and b_2 in terms of the others.

(c) If the area of a trapezoid is 85.8 sq in. and the bases are 4.6 and 3.2 in., respectively, what is its height?

(d) If the area of a trapezoid is 240 sq cm, its height is 12 cm, and one base is 27 cm, what is the other base?

6. The volume of a rectangular solid is given by $V = lwh$ where l denotes length, w denotes width, and h denotes height.

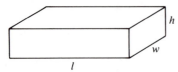

    (a) What is the volume of a box 10.4 cm long, 6.2 cm wide, and 3.6 cm high?

    (b) Solve the formula for l, for w, and for h.

    (c) If the volume of a box is 78 cu cm, its length is 5 cm, and its width is 3 cm, what is its height?

7. Solve the equation $6x + 12y = 8$ for x in terms of y. If $y = 3$, what is x?

8. Solve the equation $2x + 4y = 3$ for y in terms of x. If $x = 7$, what is y?

9. Solve the equation $11x + 13y = 5$ for x in terms of y and for y in terms of x. If $x = 5$, what is y? If $y = 11$, what is x?

10. Solve the equation $5x - 7y = 3$ for x in terms of y and for y in terms of x. If $x = 7$, what is y? If $y = 10$, what is x?

11. The area of a circle is given by the formula $A = \frac{1}{2}Cr$ where C denotes the circumference and r the radius. Express the circumference in terms of the radius and the area. Express the radius in terms of the area and the circumference.

12. The surface area of a cylinder is given by $A = C(r + h)$ where C is the circumference, h is the height, and r is the radius. Solve this formula for C, r, and h.

13. Express the height of a parallelogram in terms of the area and the base.

14. Solve the equation $ax + b = c$ for x in terms of a, b, and c.

15. Solve the equation $ax + by = c$ for y in terms of x, a, b, and c. Solve it for x in terms of y, a, b, and c.

16. Solve the equation $x + 4y + 8z = 9$ for x in terms of y and z. Solve it for y in terms of x and z. Solve it for z in terms of x and y.

17. Solve the equation $2x - 6y + 10z = 11$ for x in terms of y and z. Solve it for y in terms of x and z. Solve it for z in terms of x and y.

18. Solve the equation $ax + by + cz = d$ for each of the variables x, y, and z in terms of the others and a, b, c, and d.

4. Systems of Equations in Two Variables

Thus far we have considered only equations involving one variable, numerals, and the operations of addition, multiplication, subtraction, and division. In no case was the variable multiplied by itself nor did any expressions involving the variable divide any other expressions. All such equations can be transformed to equations of the general form

$$ax + b = 0$$

where a and b may be any numbers. Equations of this form are called *linear* equations in one variable. The origin of the word "linear" will become apparent when we study graphs of equations.

Example 1

Consider now the equation

$$2x + 3y = 8$$

A solution for this equation would be an ordered pair of numbers which when substituted for x and y would yield a true statement. If we let $x = 4$ and $y = 0$ we have a solution. This equation has other solutions, among them being

$$
\begin{array}{llll}
x = 2 & \text{and} & y = \frac{4}{3} & \text{since } 2(2) + 3(\frac{4}{3}) = 8 \\
x = \frac{5}{2} & \text{and} & y = 1 & \text{since } 2(\frac{5}{2}) + 3(1) = 8 \\
x = 0 & \text{and} & y = \frac{8}{3} & \text{since } 2(0) + 3(\frac{8}{3}) = 8
\end{array}
$$

In fact, each number we substitute for x will yield a linear equation in the one variable y and this equation will have a solution. Thus the equation

$$2x + 3y = 8$$

actually has infinitely many solutions. Realize, however, that not all ordered pairs of numbers are solutions, for example, $x = 5$ and $y = 10$ is not a solution to this equation. $\square$

Frequently when we have an equation in two variables, the problem is not just to find solutions of this equation but to find solutions which are also solutions of another equation.

Example 2

Consider now the equation above along with a second equation:

$$2x + 3y = 8$$
$$x + 4y = -1$$

Then $x = 7$ and $y = -2$ is a solution to both of these equations since

$$2(7) + 3(-2) = 8$$

and

$$7 + 4(-2) = -1 \quad \square$$

A pair of equations in two variables is usually called a *system* of two equations in two unknowns or two variables. To *solve* a system is to determine all solutions of the system, that is, all ordered pairs that are solutions of both equations. Our substitution principle for equality enables us to add, subtract, or multiply corresponding sides of equations and obtain new equations since if $A = B$ and $P = Q$, it follows that

$$A + P = B + Q, \quad A - P = B - Q, \quad \text{and} \quad AP = BQ$$

Furthermore, if $P \neq 0$, then $A/P = B/Q$. We seldom use the division technique because we must restrict ourselves to the case where the numbers are nonzero.

Example 3

Solve the system

$$2x + 3y = 9$$
$$2x + 2y = 4$$

If we subtract the left side of the second equation from the left side of the first and the right side of the second from the right side of the first, we obtain

$$(2x + 3y) - (2x + 2y) = 9 - 4$$

and by arithmetic we have

$$y = 5$$

Thus the only possible value for y is 5. When we substitute 5 for y in the first equation we obtain

$$2x + 15 = 9$$
$$2x = -6$$
$$x = -3$$

Checking in both equations shows that $x = -3$ and $y = 5$ is, in fact, a solution.

$$2(-3) + 3(5) = 9$$
$$2(-3) + 2(5) = 4 \quad \square$$

We were able to solve the system in Example 3 quite simply because $2x$ was a term in both equations and subtracting yielded an equation in the one variable y.

Example 4

Solve the system

$$2x + 5y = 5$$
$$x + 2y = 3$$

In this case, subtraction would not yield an equation in one variable. Suppose, however, that we first multiply both sides of the second equation by 2. We now have

$$2x + 5y = 5$$
$$2x + 4y = 6$$

Subtraction yields

$y = -1$

Substituting -1 for y in the second equation in our original system yields

$$x + (-2) = 3$$
$$x = 5$$

Checking shows that $x = 5$, $y = -1$ is a solution.

$$2(5) + 5(-1) = 5$$
$$5 + 2(-1) = 3 \quad \square$$

Example 5

Solve the system

$$2x + 5y = 9$$
$$3x + 2y = 4$$

In this case we could multiply both sides of the second equation by $\frac{2}{3}$ and then subtract and proceed as before. It would be computationally simpler, however, to multiply both sides of the first equation by 3, both sides of the second by 2, and then subtract.

$$6x + 15y = 27$$
$$6x + 4y = 8$$
$$11y = 19$$
$$y = \frac{19}{11}$$

We could now substitute 19/11 for y in one of the original equations and then solve for x. An alternative method is to now choose multipliers so that subtracting yields an equation in the one variable x. If we use 2 as a multiplier in the first case and 5 in the second we obtain

$$2(2x + 5y) = 2 \cdot 9$$
$$5(3x + 2y) = 5 \cdot 4$$

or

$$4x + 10y = 18$$
$$15x + 10y = 20$$

Subtracting yields

$$-11x = -2 \quad \text{and} \quad x = \frac{2}{11}$$

Checking shows that

$$2\left(\frac{2}{11}\right) + 5\left(\frac{19}{11}\right) = \frac{4 + 95}{11} = \frac{99}{11} = 9$$

and

$$3\left(\frac{2}{11}\right) + 2\left(\frac{19}{11}\right) = \frac{6 + 38}{11} = \frac{44}{11} = 4$$

Thus $x = 2/11$, $y = 19/11$ is the solution to the system. $\square$

A third alternative is first to solve one of the equations for one variable in terms of the other. Then substitute this expression in the other equation to obtain an equation in one variable which we then proceed to solve.

Example 6

Solve

$$x + 2y = 5$$
$$2x + 3y = 9$$

Solving the first equation for x in terms of y yields

$$x = 5 - 2y$$

Substituting $5 - 2y$ for x in the second equation yields

$$2(5 - 2y) + 3y = 9$$

Then

$$10 - 4y + 3y = 9$$
$$10 - y = 9$$
$$y = 1$$

Substituting 1 for y in $x = 5 - 2y$ yields

$$x = 5 - 2 \cdot 1 = 3$$

Checking verifies that

$$3 + 2 \cdot 1 = 5$$
$$2 \cdot 3 + 3 \cdot 1 = 6 + 3 = 9$$

Thus $x = 3$, $y = 1$ is a solution. $\square$

The process of choosing multipliers and adding or subtracting in order to obtain an equation in one variable is the most automatic technique but is not always the simplest. In practice, the most common technique involves using substitution to find the second number if the computation is not too messy.

All of the equations shown thus far had solutions and all the systems of equations did also. The student should realize, however, that it is possible to write equations with no solution just as can use pronouns to write sentences that are never true.

Example 7

The equation

$$x = x + 1$$

has no solution since

$$x < x + 1$$

for all numbers x. ☐

Example 8

The system of equations

$$2x + 3y = 5$$
$$2x + 3y = 8$$

can have no solution because the number $2x + 3y$ cannot equal both 5 and 8. If we subtract with this system we obtain $0 = -3$ which we know is false. ☐

Example 9

The system

$$2x + 3y = 4$$
$$4x + 6y = 8$$

has more than one solution because we have, essentially, only one equation. That is, although the equations are different we can obtain the second by multiplying both sides of the first by 2. We could also obtain the first by multiplying both sides of the second by $\frac{1}{2}$. Any solution of one is a solution of the other and vice versa. ☐

Thus we see that with a system of two equations we have three possibilities:

1. no solution
2. exactly one solution
3. an infinite number of solutions

We will see in Chapter 6 that systems of the type

$$ax + by = c$$
$$px + qy = r$$

have more than one solution only when every solution of one is a solution of the other.

Before proceeding to the exercises, study the following examples, justify each step, and check the solutions.

Example 10

Solve the system

$$x + 3y = 5$$
$$2x - y = 4$$

$$\begin{array}{ll} 2x + 6y = 10 & x + 3y = 5 \\ 2x - y = 4 & 6x - 3y = 12 \end{array}$$

$$\begin{array}{ll} 7y = 6 & 7x = 17 \\ y = \frac{6}{7} & x = \frac{17}{7} \end{array}$$

Example 11

Solve the system

$$.1x - .3y = 7.2$$
$$2x + y = 4$$

$$\begin{array}{ll} x - 3y = 72 \\ 2x + y = 4 & y = 4 - 2x \end{array}$$

$$x - 3(4 - 2x) = 72$$
$$x - 12 + 6x = 72$$
$$7x = 84$$
$$x = 12$$
$$y = 4 - 2(12)$$
$$y = -20 \ \square$$

EXERCISES

1. Solve the following systems by choosing multipliers and adding or subtracting.

(a) $3x - 2y = 8$
  $x + y = 6$

(b) $7x + 5y = 16$
  $3x - 2y = 11$

(c) $1.3x - 5.2y = -13$
  $7x + 2y = 90$

(d) $\frac{1}{2}x - \frac{1}{3}y = \frac{3}{2}$
  $\frac{2}{3}x - \frac{7}{2}y = \frac{9}{4}$

(e) $\frac{2}{3}x + \frac{7}{6}y = 10$
  $\frac{3}{2}x - \frac{2}{3}y = -17$

(f) $16x - 5y = .45$
  $6x + 7y = 79$

(g) $9x + 6y = 5$
  $2x - y = 19$

(h) $.43x - 10.2y = 10.96$
  $8.4x - 16y = 32.8$

(i) $\frac{9}{2}x + \frac{4}{7}y = \frac{11}{3}$
  $\frac{6}{7}x + \frac{5}{4}y = \frac{2}{3}$

(j) $5x - 7y = 42$
  $6x + 8y = 19$

2. Solve the following systems using addition and multiplication to solve for one variable and substitution to solve for the other.

(a) $3x - 2y = 5$
$2x - y = 6$

(b) $x + 3y = -5$
$4x + 5y = 1$

(c) $2x + 13y = 4$
$x + 5y = 11$

(d) $\frac{4}{5}x + \frac{2}{3}y = 22$
$x - \frac{3}{4}y = 10$

(e) $1.3x - 7.2y = -.98$
$4.3x + y = 9.2$

(f) $\frac{1}{7}x + \frac{1}{5}y = 1$
$x - y = 2$

(g) $3x - 7y = -2$
$5x + 3y = 26$

(h) $2x + 5y = 19$
$x + 2y = 14$

(i) $4.8x - 3y = 8$
$3.2x + 4y = 9$

(j) $80x + 92y = 113$
$3x + 2y = 9$

3. Solve the following systems.

(a) $15x - 17y = 9$
$2x - 3y = -1$

(b) $453x + 241y = 99$
$x - y = -7$

(c) $\frac{1}{4}x + \frac{1}{5}y = 1$
$\frac{1}{3}x - \frac{2}{3}y = 1$

(d) $5.1x + 1.2y = -5$
$x + .1y = 6$

(e) $4x - 7y = 11$
$8x + 3y = 5$

(f) $5x + 2y = -4$
$x - 3y = 2$

(g) $.013x + .09y = -4.2$
$20x - 13y = 91$

(h) $6x + 2y = 5$
$x - y = 9$

(i) $3x - 7y = 4$
$5x + 3y = 19$

(j) $x + 3y = 91$
$2x + y = 97$

(k) $4.3x + 3.9y = 2.6$
$8x - y = 3$

(l) $14x + 17y = 11$
$5x - 3y = 16$

(m) $86x + 71y = 20$
$2x + y = 5$

(n) $81x + 19y = 17$
$57x + 11y = 23$

4. Solve the following systems of equations.

(a) $2x + 3y = 17$
$5x + 7y = 31$

(b) $2x + 3y = 11$
$5x + 7y = 19$

(c) $2x + 3y = 13$
$5x + 7y = 23$

(d) $2x + 3y = p$ for arbitrary numbers p and q
$5x + 7y = q$

(e) $ax + by = 11$ for arbitrary numbers $a, b, c,$ and d, and with $ad - bc \neq 0$
$cx + dy = 19$

*(f) $ax + by = p$ for arbitrary numbers $a, b, c, d, p,$ and q and with
$cx + dy = q$ $ad - bc \neq 0$

*5. Use your solutions to 4(f) to solve the following systems.

(a) $2x + 7y = 8$
$3x + 10y = 4$

(b) $8x - 5y = 4$
$3x + 2y = 9$

(c) $6x + 5y = 11$
$7x - 2y = 13$

(d) $3x - 4y = 2$
$5x + 2y = 6$

5. Systems of Equations in Three or More Variables

The techniques involved in solving systems of two equations in two unknowns generalize to three equations in three unknowns and, in fact, to n equations in n unknowns.

Example 1

Solve

$$x + 2y + 3z = -4$$
$$2x - y - z = 5$$
$$3x + 2y - 2z = 20$$

If we multiply both sides of the first equation by 2, then the first and second equations yield

$$2x + 4y + 6z = -8$$
$$2x - y - z = 5$$

Subtracting yields

$$5y + 7z = -13$$

If we multiply both sides of the first equation by 3, the first and third equations yield

$$3x + 6y + 9z = -12$$
$$3x + 2y - 2z = 20$$

Subtracting yields

$$4y + 11z = -32$$

Now we have two equations in two unknowns, y and z:

$$5y + 7z = -13$$
$$4y + 11z = -32$$

We now proceed as in the previous section when solving systems of two equations. Using 4 and 5 as multipliers we have

$$20y + 28z = -52$$
$$20y + 55z = -160$$

Subtracting yields

$$-27z = 108 \quad \text{and} \quad z = -4$$

Substituting into

$$5y + 7z = -13$$

we have

$$5y + 7(-4) = -13$$
$$5y - 28 = -13$$
$$5y = 15$$
$$y = 3$$

Now substituting into the first equation we have

$$x + 2(3) + 3(-4) = -4$$
$$x + 6 - 12 = -4$$
$$x - 6 = -4$$
$$x = 2$$

Thus $x = 2$, $y = 3$, and $z = -4$ should be a solution to the system. Checking shows that it is:

$$2 + 6 - 12 = -4$$
$$4 - 3 + 4 = 5$$
$$6 + 6 + 8 = 20 \ \square$$

As with systems of two equations, we could solve for each unknown by eliminating variables but in most cases, the computation is simpler if we use substitution after values have been determined for one or two variables. Systems of more than three equations are solved by similar techniques.

Example 2

Solve the system

$$2x + y + 2z + w = 6$$
$$3x - y - 4z - 2w = 13$$
$$x + 2y - 3z - 4w = -16$$
$$5x + 3y + z - 3w = -6$$

Adding corresponding sides of the first two equations yields

$$5x - 2z - w = 19$$

If we multiply both sides of the second equation by 2, then the second and third equations become

$$6x - 2y - 8z - 4w = 26$$
$$x + 2y - 3z - 4w = -16$$

Adding yields

$$7x - 11z - 8w = 10$$

Multiplying both sides of the second equation by 3 and adding the sides to the corresponding sides of the fourth equation yields

$$14x - 11z - 9w = 33$$

We now have a system of three equations in three variables, x, z, and w and we can proceed as before:

$$5x - 2z - w = 19$$
$$7x - 11z - 8w = 10$$
$$14x - 11z - 9w = 33$$

As always, we have a choice of which variable to eliminate next. If we subtract the sides of the second equation from the corresponding sides of the third we have

$$7x - w = 23$$

If we multiply both sides of the first equation by 11 and both sides of the second by 2 we have

$$55x - 22z - 11w = 209$$
$$14x - 22z - 16w = 20$$

Subtracting yields

$$41x + 5w = 189$$

We now have a system of two equations in two unknowns:

$$7x - w = 23$$
$$41x + 5w = 189$$

Multiplying both sides of the first equation by 5 yields

$$35x - 5w = 115$$
$$41x + 5w = 189$$

Adding yields

$$76x = 304$$

and

$$x = 4$$

Our system at this point is

$$x = 4$$
$$7x - w = 23$$
$$5x - 2z - w = 19$$
$$2x + y + 2z + w = 6$$

Substituting 4 for x in the second equation we have

$$28 - w = 23$$
$$w = 5$$

Substituting 4 for x and 5 for w in the third yields

$$20 - 2z - 5 = 19$$
$$-2z = 4$$
$$z = -2$$

Substituting 4 for x, 5 for w, and -2 for z in the last yields

$$8 + y - 4 + 5 = 6$$
$$y = -3$$

Thus $x = 4$, $y = -3$, $z = -2$, $w = 5$ should be our solution. We substitute this in the original system and see that it is.

$$2(4) + (-3) + 2(-2) + 5 = 6$$
$$3(4) - (-3) - 4(-2) - 2(5) = 13$$
$$4 + 2(-3) - 3(-2) - 4(5) = -16$$
$$5(4) + 3(-3) + (-2) - 3(5) = -6 \quad \square$$

Systems of five equations in five unknowns are solved in a similar fashion. We eliminate one variable to obtain a system of four equations in four unknowns and proceed as before. Remember, however, that not all systems have solutions.

EXERCISES

Solve the following systems of equations.

1. $x + 2y - 3z = 10$
   $2x - y + 5z = -15$
   $4x + 3y - 2z = 5$

2. $5x - 3y + 2z = 30$
   $6x + y - z = 35$
   $2x + 3y + z = 17$

3. $x + 2y + 3z = 20$
   $2x - y + 2z = 5$
   $4x + 3y + 8z = 15$

4. $x + 2y - 4z = 7$
   $3x + 5y - 2z = 0$
   $x + y - z = 14$

5. $.5x + .3y + z = 4$
   $.3x - 2.1y + z = .8$
   $6x + 7y - 1.1z = 3$

6. $2x + 3y - z = 5$
   $3x - y + 2z = 3$
   $9x + 8y + z = 10$

7. $3x + 4y + z = 3$
   $4x - y + 2z = 20$
   $5x + 8y - z = -11$

8. $2x + 3y + 5z = -1$
   $4x - y + 2z = 9$
   $3x + y + z = 14$

9. $\frac{1}{3}x - \frac{2}{3}y + \frac{1}{4}z = \frac{1}{8}$
   $\frac{2}{5}x - \frac{1}{4}y + \frac{3}{2}z = \frac{3}{5}$
   $\frac{1}{2}x + \frac{1}{3}y - \frac{1}{6}z = \frac{5}{3}$

10. $\frac{3}{5}x + \frac{4}{5}y + \frac{1}{5}z = \frac{4}{5}$
    $\frac{4}{3}x - \frac{1}{3}y + \frac{2}{3}z = \frac{10}{3}$
    $\frac{5}{8}x + y - \frac{1}{8}z = \frac{1}{4}$

11. $2x + 3y = -1$
    $3x + 4z = 20$
    $5y + 6z = -3$

12. $2x + 3y = -11$
    $3x + 4z = 18$
    $5y + 6z = -7$

13. $2x + 3y = p$
    $3x + 4z = q$ for arbitrary
    $5y + 6z = r$ numbers p, q, r

14. Solve the general system
    $ax + by = p$
    $ex + fz = q$
    $gx + kz = r$

15. $2x + 5y + 7z = 9$
    $3x + 11y + 13z = 9$
    $x + 3y + 2z = 2$
17. $2x + 5y + 7z = 13$
    $3x + 11y + 13z = 5$
    $x + 3y + 2z = -4$
*19. Solve the general system
    $ax + by + cz = p$
    $dx + ey + fz = q$
    $gx + hy + kz = r$
    (make the needed assumption as
    to certain numbers being nonzero)
21. $2x + 3y - z + w = 5$
    $3x - y + 2z + 3w = 22$
    $x + 2y + 3z - w = 33$
    $4x - 4y + z + 2w = 10$

16. $2x + 5y + 7z = 8$
    $3x + 11y + 13z = 7$
    $x + 3y + 2z = 5$
18. $2x + 5y + 7z = p$ for arbi-
    $3x + 11y + 13z = q$ trary num-
    $x + 3y + 2z = r$ bers p, q, r
*20. Use your result from Exercise 19
    to solve the system:
    $2x + 3y + 4z = 5$
    $3x + 2y + 5z = 11$
    $5x + 6y + 11z = 4$
22. $x + y + z + w = 13$
    $2x + 3y - 4z + 2w = -6$
    $3x + y - z + 3w = 20$
    $x + 2y + 3z + w = 21$

6. Solving Linear Inequalities

Thus far in this chapter we have concerned ourselves with solving linear equations and systems of linear equations. In Chapter 1, we pointed out that comparability was one of the important properties of numbers. Sometimes we don't care exactly what the numbers are but how they are related. For example, in a football game the scores themselves are not the important thing, all we care is that our team's score is greater than the other team's score. If we are driving an automobile on a trip, we do not want our tank to contain just enough gasoline but rather we make sure that it contains at least enough. The fish and game department in your state sets limits on certain species and requires that a sportsman take no more than these limits; he certainly can take less.

Since an equation or equality is a sentence of the form $A = B$, we might expect an inequality to be a statement of the form $A \neq B$. Since, however, any two numbers are comparable, if A and B are numbers and $A \neq B$, then either $A < B$ or $B < A$. For our purposes, in the study of inequalities we will be concerned only with statements of the form

$$A < B, \quad A \leq B, \quad A \geq B, \quad \text{or} \quad A > B$$

If an inequality contains one variable, say x, to solve the inequality means to determine all numbers which when substituted for x yield a true statement. In general, we can expect an inequality to have more solutions than is expected of an equation. For example, consider the following two sentences:

$$2x < 6 \qquad 2x = 6$$

Any number smaller than 3 will be a solution of the inequality whereas 3 is the only solution of the equation.

The techniques for solving inequalitites are based on three principles previously discussed in Chapter 1.

A. If a, b, and c are numbers such that $a < b$, then $a + c < b + c$ and $a - c < b - c$.

B. If a, b, and c are numbers such that $a < b$ and c is positive, then $ac < bc$ and $a/c < b/c$.

C. If a, b, and c are numbers such that $a < b$ and c is negative, then $bc < ac$ and $b/c < a/c$.

Analogous statements hold if we replace " $<$ " in these statements by " $>$," " $\leq$," or " $\geq$." These statements can be condensed as follows:

(i) If we add or subtract the same number to or from both sides of an inequality, we obtain an inequality in the *same* direction.

(ii) If we multiply or divide both sides of an inequality by a *positive* number, we obtain an inequality in the *same* direction.

(iii) If we multiply or divide both sides of an inequality by a *negative* number, we obtain an inequality in the *opposite* direction.

Example 1

$x + 3 < 7$	Given
$(x + 3) - 3 < 7 - 3$	Subtracting 3 from both sides yields an inequality in the same direction.
$x < 4$	Arithmetic $\square$

The solution to an inequality is a set of numbers. We could say now that the solution to the inequality $x + 3 < 7$ is the set $\{x : x < 4\}$. This is a lot of extraneous writing and really does not convey much meaning. The set of solutions to the original inequality is the set of numbers satisfying $x < 4$ and we will consider the statement $x < 4$ as sufficient in describing the solutions.

Example 2

$y - 4 < -5$	Given
$(y - 4) + 4 < (-5) + 4$	Adding 4 to both sides yields an
$y < -1$	inequality in the same direction.

The statement $y < -1$ is sufficient to describe all solutions. $\square$

Example 3

$$2z \geq 7$$

$$\frac{2z}{2} \geq \frac{7}{2}$$

$$z \geq \frac{7}{2} \; \square$$

Dividing both sides by the positive number 2 yields an inequality in the same direction.

Example 4

$$-3x > 5$$

$$\frac{-3x}{-3} < \frac{-5}{3}$$

$$x < \frac{-5}{3} \; \square$$

Dividing both sides by -3 yields an inequality in the opposite direction.

As in solving equations, we may need to apply these techniques several times before obtaining a solution.

Example 5

$$4x + 7 \geq 9$$
$$(4x + 7) - 7 \geq 9 - 7$$
$$4x \geq 2$$

Subtracting 7 from both sides preserves the inequality.

$$\frac{4x}{4} \geq \frac{2}{4}$$

$$x \geq \frac{1}{2} \; \square$$

Dividing both sides by 4 preserves the inequality.

Example 6

$$5x - 3 < 7x + 5$$
$$(5x - 3) + 3 < (7x + 5) + 3$$
$$5x < 7x + 8$$
$$5x - 7x < 7x + 8 - 7x$$
$$-2x < 8$$

Adding 3 to both sides preserves the inequality.
Subtracting $7x$ from both sides preserves the inequality.

$$\frac{-2x}{-2} > \frac{-8}{2}$$

$$x > -4 \; \square$$

Dividing both sides by -2 reverses the direction of the inequality.

Example 7

$$.5y + 6 < .4y - 2$$
$$(.5y + 6) - 6 < .4y - 2 - 6$$
$$.5y < .4y - 8$$
$$.5y - .4y < .4y - 8 - .4y$$
$$.1y < -8$$
$$y < -80 \ \square$$

(The reader should provide reasons for each step.)

The solution set of an inequality may be easier to grasp if we draw a graph or picture of the set on a number line. This is done by drawing a number line in the usual fashion and then indicating the solution set by a heavier line. Figure 2-1 provides several examples.

Figure 2-1

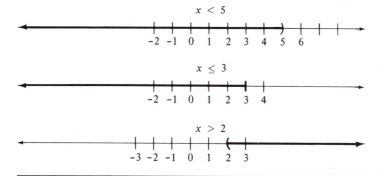

Notice that we indicate that the end point is not included by using a form such as "─)" or "(─" and that it is included is indicated by a form such as "─|" or "|─."

All the inequalities considered thus far are of the form

$$ax + b < cx + d$$
$$ax + b \le cx + d$$
$$ax + b > cx + d$$
$$ax + b \ge cx + d$$

Such inequalities can be solved without multiplying or dividing by expressions containing the variable, that is, all multiplication and division of both sides uses only constants and therefore we know which of principles (ii) and (iii) we are applying. Later on we will see situations where we must consider both possibilities.

Frequently we have two or more inequalities and we wish to find all numbers that satisfy both of them or either of them.

Figure 2-2

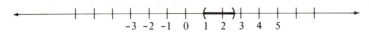

Example 8

Find all numbers x such that

$$2x + 3 < 8 \quad and \quad 3x - 1 > 2$$

First we solve both inequalities:

$$2x + 3 < 8 \qquad 3x - 1 > 2$$
$$2x < 5 \qquad\quad 3x > 3$$
$$x < \tfrac{5}{2} \qquad\quad x > 1 \quad \text{that is, } 1 < x$$

Thus the set of all solutions to both inequalities is the set of all numbers x such that

$$1 < x \quad and \quad x < \tfrac{5}{2}$$

A simpler way of denoting this is by

$$1 < x < \tfrac{5}{2}$$

Notice that the solution set for the system is the *intersection* of the solution sets for the two inequalities. The graph is given in Figure 2-2. ☐

Example 9

Find the solution set of

$$5x + 2 < 12 \quad and \quad 2x - 4 > 2$$

Solving both inequalities yields two solution sets:

$$5x + 2 < 12 \qquad 2x - 4 > 2$$
$$5x < 10 \qquad\quad 2x > 6$$
$$x < 2 \qquad\quad\; x > 3$$

In order for a number to satisfy both inequalities, it would have to be less than 2 *and* greater than 3 and this is impossible. Thus this system has no solutions, that is, it has an empty solution set. ☐

Example 10

Find all numbers satisfying

$$4x + 2 < 14 \quad and \quad 3x - 1 < 14$$

Figure 2-3

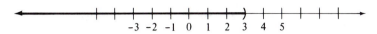

Solving both inequalities yields the two solution sets:

$$4x + 2 < 14 \qquad 3x - 1 < 14$$
$$4x < 12 \qquad\quad 3x < 15$$
$$x < 3 \qquad\quad\; x < 5$$

The solution set for the system is all numbers that are less than 3 *and* less than 5. These are just the numbers less than 3. Thus $\{x: x < 3\}$ is the solution set for the system. The graph is given in Figure 2-3. □

Example 11

Find all numbers satisfying

$$3x + 1 < 5 \quad or \quad 2x + 1 > 7$$

Solving both inequalities yields the two solution sets:

$$3x < 4 \qquad\qquad 2x > 6$$
$$x < \tfrac{4}{3} \qquad\qquad x > 3$$
$$\{x: x < \tfrac{4}{3}\} \qquad \{x: 3 < x\}$$

The desired solution set is the union of these two sets:

$$\{x: x < \tfrac{4}{3} \quad or \quad x > 3\}$$

The graph is given in Figure 2-4. □

Figure 2-4

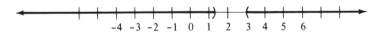

Example 12

Find all numbers satisfying

$$5 < 2x + 4 \quad or \quad 2x + 4 < -5$$

Figure 2-5

Solving both inequalities yields

$$1 < 2x \qquad 2x < -9$$
$$\tfrac{1}{2} < x \qquad x < -\tfrac{9}{2}$$

The desired solution set is all numbers satisfying either inequality, that is,

$$\{x : \tfrac{1}{2} < x \quad or \quad x < -\tfrac{9}{2}\}$$

The graph is given in Figure 2-5. □

Example 13

Find all numbers satisfying

$$7 \le 3x - 2 \quad or \quad 3x - 2 \le -7$$

Solving both inequalities yields

$$9 \le 3x \qquad 3x \le -5$$
$$3 \le x \qquad x \le -\tfrac{5}{3}$$

The solution set is

$$\{x : 3 \le x \quad or \quad x \le -\tfrac{5}{3}\}$$

The graph is given in Figure 2-6. □

Figure 2-6

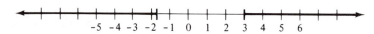

Example 14

Find all numbers satisfying

$$3x + 7 \ge 5x + 2 \quad and \quad 6x - 4 < 8x + 3$$

Solving both inequalities yields

$$5 \ge 2x \quad and \quad -7 < 2x$$
$$\tfrac{5}{2} \ge x \quad and \quad -\tfrac{7}{2} < x$$

Figure 2-7

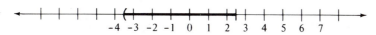

that is,

$x \leq \frac{5}{2}$ and $-\frac{7}{2} < x$

Thus the solution set is

$\{x: -\frac{7}{2} < x \leq \frac{5}{2}\}$

The graph is given in Figure 2-7. □

Example 15

Find all numbers satisfying

$5x + 11 < 2x + 3$ *or* $7x - 4 \geq 5x + 11$

Solving both inequalities yields

$3x < -8$ or $2x \geq 15$
$x < -\frac{8}{3}$ or $x \geq \frac{15}{2}$

The solution set is

$\{x: x < -\frac{8}{3}$ *or* $x \geq \frac{15}{2}\}$

The graph is given in Figure 2-8. □

Figure 2-8

Example 16

Solve

$|x| > 4$

By the definition of absolute value, $|x| = x$ or $|x| = -x$. Thus $|x| > 4$

Figure 2-9

$$|x| > 4$$

holds if $x > 4$ or $-x > 4$, that is, if $x > 4$ or $x < -4$. The graph of the system $|x| > 4$ is given in Figure 2-9. $\square$

Example 17

Solve

$$|x| < 5$$

Since

$$x \le |x| \quad \text{and} \quad -x \le |x|$$

if $|x| < 5$, then $x < 5$ and $-x < 5$. But if $x < 5$, and $-x < 5$, then $x < 5$ and $x > -5$. Thus if $|x| < 5$, $x < 5$ and $x > -5$. The graph of $|x| < 5$ is given in Figure 2-10. $\square$

Figure 2-10

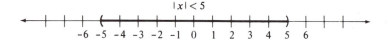

$$|x| < 5$$

In general, if a is a positive number, then $|y| > a$ if and only if $y > a$ or $y < -a$. Thus the inequality $|y| > a$ is equivalent to the system of inequalities $y > a$ or $y < -a$.

Similarly, if a is a positive number, then $|y| < a$ if and only if $-a < y$ and $y < a$, that is, $-a < y < a$. Thus the inequality $|y| < a$ is equivalent to the system $-a < y$ and $y < a$. (See Figure 2-11.)

Figure 2-11

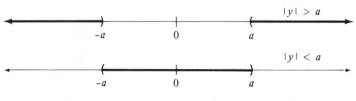

$$|y| > a$$

$$|y| < a$$

Figure 2-12

Example 18

Solve

$$|3x - 1| < 6$$

This inequality is equivalent to the system of inequalities

$$-6 < 3x - 1 \quad and \quad 3x - 1 < 6$$

Solving yields

$$-5 < 3x \quad and \quad 3x < 7$$
$$-\tfrac{5}{3} < x \quad and \quad x < \tfrac{7}{3}$$

Thus the solution set is

$$\{x: -\tfrac{5}{3} < x \text{ and } x < \tfrac{7}{3}\}$$

This can be written simply as

$$\{x: -\tfrac{5}{3} < x < \tfrac{7}{3}\} \quad \text{or} \quad -\tfrac{5}{3} < x < \tfrac{7}{3}$$

The graph is given in Figure 2-12. □

Example 19

Solve

$$|2x + 3| > 7$$

This inequality is equivalent to the system

$$7 < 2x + 3 \quad or \quad 2x + 3 < -7$$

Solving yields

$$4 < 2x \quad or \quad 2x < -10$$
$$2 < x \quad or \quad x < -5$$

The solution set is

$$\{x: 2 < x \text{ or } x < -5\}$$

The graph is given in Figure 2-13. □

Figure 2-13

Figure 2-14

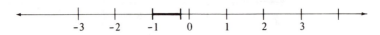

Notice that in Example 18 the solution set is the intersection of the two solution sets whereas in Example 19 the solution set is the corresponding union.

Example 20

Solve $|5x + 3| \leq 2$.

$$-2 \leq 5x + 3 \quad and \quad 5x + 3 \leq 2$$
$$-5 \leq 5x \quad\quad and \quad\quad 5x \leq -1$$
$$-1 \leq x \quad\quad and \quad\quad x \leq -\tfrac{1}{5}$$

The solution set is

$$\{x: -1 \leq x \leq -\tfrac{1}{5}\}$$

The graph is given in Figure 2-14. $\square$

Example 21

Solve $|5x - 1| \geq 11$.

$$5x - 1 \geq 11 \quad or \quad 5x - 1 \leq -11$$
$$5x \geq 12 \quad or \quad\quad 5x \leq -10$$
$$x \geq \tfrac{12}{5} \quad or \quad\quad x \leq -2$$

The solution set is

$$\{x: x \geq \tfrac{12}{5} \text{ or } x \leq -2\}$$

The graph is given in Figure 2-15. $\square$

Figure 2-15

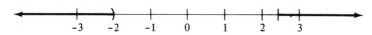

EXERCISES

1. Solve the following inequalities. Write out reasons for each step. Draw graphs of the solution sets.

(a) $-2x + 5 < 8$ (b) $4x - 7 < 9$
(c) $.3x + 2 < .8$ (d) $3x + 2 \geq 4x - 7$
(e) $5x - 2 \leq 13x + 9$ (f) $2x + 5 < 6x + 19$
(g) $15x + 3 \geq 11x - 7$ (h) $\frac{1}{3}x + \frac{1}{6} > \frac{4}{3}x + \frac{9}{7}$
(i) $\frac{4}{3}y - \frac{2}{3} \leq \frac{1}{6}y + \frac{5}{3}$ (j) $4(x - 5) < 3(x + 2)$
(k) $7(6 - 5x) \geq 4(3x + 2)$ (l) $9(4x - 3) < 8(4 - 3x)$

2. Solve the following inequalities. Graph the solutions.
    (a) $\frac{1}{3}(x + 1) < \frac{2}{3}(x + 2)$ (b) $5(x + 3) \geq 16(x - 2)$
    (c) $6(2x + 1) < 8(x + 3)$ (d) $.6(x - 3) \leq .7(x + 2)$
    (e) $17(2x - 1) < 16(3x + 1)$ (f) $\frac{4}{5}(2x - 3) \geq 4(\frac{1}{3}x - 2)$
    (g) $.9(.4x - 3) < .6(.8x - 7.1) + 4.3$ (h) $\left(-\frac{2}{3}\right)(4x - \frac{1}{9}) \leq \left(\frac{4}{5}\right)(3x - 7)$
    (i) $\left(\frac{2}{3}\right)(\frac{4}{3}x - \frac{2}{19}) < \left(-\frac{4}{7}\right)(\frac{3}{5}x + \frac{4}{19})$ (j) $6(x + 3) + 7(2x - 1) > 8(3x - 7)$

3. Solve the following systems of inequalities. Graph the solutions.
    (a) $-2x + 5 < 8$ and $4x - 7 < 9$
    (b) $3x + 2 \geq 4x - 7$ and $5x - 2 \leq 13x + 9$
    (c) $2x + 3 < 7$ and $3x + 1 > 2$
    (d) $7x + 3 \leq 19$ and $3x - 1 < 4$
    (e) $4x + 1 > 2$ or $3x - 2 < 1$
    (f) $6x - 5 < 3$ and $-3 < 6x - 5$
    (g) $13x - 19 \geq 5$ or $13x - 19 \leq -5$
    (h) $1.5x - 7 \leq 3$ or $9.2x + 4 \geq 9$
    (i) $2x + 3 < 3x - 2$ and $5x + 5 < 8x + 2$
    (j) $6x - 7 > 4x - 5$ and $3x + 2 < 6x - 7$

4. Solve the following systems of inequalities. Graph the solution sets.
    (a) $3x - 1 < 18$ and $3x - 1 > -18$
    (b) $4x + 3 \leq 12$ and $4x + 3 \geq -12$
    (c) $2x + 1 > 16$ or $2x + 1 < -16$
    (d) $5x + 2 < 9$ and $5x + 2 > -9$
    (e) $6x + 3 < 7$ and $6x + 3 > -7$
    (f) $5x - 4 \geq 6$ or $5x - 4 \leq -6$
    (g) $6x - 7 < 11$ and $6x - 7 > -11$
    (h) $2x - 7 < -19$ or $2x - 7 > 19$

5. Solve the following inequalities. Graph the solution sets.
    (a) $|3x - 1| < 8$ (b) $|4x + 3| \leq 2$
    (c) $|3 - 7x| < 5$ (d) $|2x - 1| > 6$
    (e) $|3x - 2| \leq 14$ (f) $|.3x + .04| < .62$
    (g) $|.8x - .7| \geq .4$ (h) $|91x + 93| > 41$
    (i) $|\frac{2}{3}x - \frac{1}{2}| < \frac{4}{7}$ (j) $|\frac{4}{3}x + \frac{2}{3}| \geq \frac{1}{6}$

7. Problem Solving

In Sections 1–6 we developed methods for solving several types of equations, inequalities, and systems of equations. In all cases we began with certain conditions on numbers, say x or y or both, and determined just what numbers x and y might represent. General problem solving involves this same notion. Given certain conditions on quantities, what might these quantities be? For example, given that your car has traveled 450 mi in $8\frac{1}{4}$ hr, what is your average rate of speed?

Problem solving in mathematics frequently involves a translation process. The problem usually arises in a real world situation but we work in the idealized world of mathematics. For example, if our problem involves the length of a board, we translate the problem to one involving line segments (which are abstract ideas in the mathematical world), solve an analogous problem involving segments, and then translate our answer back to the physical world. Usually we are not aware of this process but every time we draw a picture and apply our geometric knowledge we are involved in such a translation. In general, then, to solve a problem using mathematics, we first need to formulate a mathematical problem that will provide an answer to the real world problem. This process of building or using mathematical models is the essence of applied mathematics. In this section we will consider a number of problems that can be solved using the algebraic techniques covered in the first five sections.

Example 1

What percent of 92 is 11?

If we translate this to symbols we might have

_____ percent of 92 = 11

We know, for instance, that 5 percent of a number means 5/100 times the number.

If we let x become the number we wish to find, we have

$$\frac{x}{100} \cdot 92 = 11$$

If we now solve this equation for x we have

$$\frac{92x}{100} = 11$$

$$92x = 11(100)$$
$$= 1100$$

$$x = \frac{1100}{92}$$

and x is approximated by 11.96. Thus 11 is approximately 11.96 percent of 92. □

When many students read this example for the first time, their first reaction is "What a lot of bother! I learned in fifth grade to determine what percent one number is of another and I didn't need algebra!"

If this was all we were going to do, they would be right and you could close the book right now. The point of this example is that any percentage problem involves the basic equation:

$$x\% \text{ of } y = z \quad \text{or} \quad \frac{x}{100} \cdot y = z$$

The fundamental percentage relationship is given by this equation and in solving any elementary percentage problem, we only need know which numbers are given and solve for the unknown.

Example 2

Suppose you own a home that you wish to sell and you need to receive net proceeds of $31,500 in order to buy the bigger home you desire. If the legal fees are $200, the real estate commission 7 percent of the selling price, and the transfer tax is 1 percent of the selling price, what must your selling price be in order to net $31,500?

Does the problem sound difficult? It isn't if you translate it into a system of equations.

Let x be the selling price of the house.
Let y be the costs of selling the house.
You require $31,500 as the net proceeds.

Since (selling price) − (costs) = net proceeds, the first equation is

$$x - y = \$31,500$$

Furthermore, the costs are:

$.07x$	for real estate commission
$.01x$	for transfer tax
$\$200$	for legal fees

Therefore

$$y = .07x + .01x + 200$$

or

$$y = .08x + 200$$

is the second equation.

Now to solve the problem, we need to solve the system

$$x - y = 31,500$$
$$y = .08x + 200$$

Substituting $.08x + 200$ for y in the first equation yields

$$x - (.08x + 200) = 31,500$$
$$.92x - 200 = 31,500$$
$$.92x = 31,700$$
$$x = \frac{31,700}{.92}$$

x is approximately 34,456.52.

If you price the house at $34,500, you would have what you need. Realize that solving this problem would not, however, determine your market price. You would price the home at or above the appraised value but would not take less than $34,456.52 because any lower price would not satisfy your needs. If you could not sell for at least this much, you would not sell. □

Example 3

Suppose you paid $240.35 for a stereo including 4.5 percent sales tax. What was the actual price of the stereo and how much tax did you pay? The basic equation is

(selling price) + (tax) = (total cost)

If we let x denote the selling price, then $.045x$ is the tax and the equation becomes

$$x + .045x = 240.35$$

Then

$$1.045x = 240.35$$
$$x = \frac{240.35}{1.045}$$
$$= 230$$

The actual price of the stereo was $230. □

The last two problems involved percentage and it is doubtful that your methods from elementary arithmetic would have solved the problems for you.

The general technique for solving problems by using algebra is:

  (i) Determine the relationships among the quantities under discussion and state them as verbal equations or inequalities.
 (ii) Express the quantities using algebraic representations.
(iii) State the relationships using the algebraic representations.
(iv) Solve the resulting equation, inequality, or system.

Example 4

Suppose your automobile runs smoothly on both leaded and nonleaded gasoline. If leaded gasoline costs 63.9¢ per gallon and yields 16 mpg and nonleaded gasoline costs 58.9¢ per gallon, how many miles per gallon must you get in order for you to benefit financially by using nonleaded gasoline?

The question is, what gas mileage on nonleaded gas assures that:

cost per mile on nonleaded gas < cost per mile on leaded gas

If we let x denote the number of miles per gallon you get on nonleaded gas, then $58.9/x$ represents the cost per mile on nonleaded gas and $63.9/16$ is the cost per mile on leaded gas.

Thus to solve the problem, we need to solve the inequality

$$\frac{58.9}{x} < \frac{63.9}{16}$$

Since the cost per mile is positive, $x > 0$ and we can multiply both sides by $16x$ and preserve the direction of the inequality. Thus

$$(58.9)(16) < 63.9x$$

and

$$\frac{(58.9)(16)}{63.9} < x$$

$$\frac{942.4}{63.9} < x$$

For practical purposes we would approximate the quotient by 14.7

$$14.7 < x$$

Thus if the nonleaded gasoline yields 14.7 or more miles per gallon, you save money. □

Example 5

Suppose you are offered two equally pleasant positions for the summer vacation. The first pays $2.75 an hour with a 10¢ raise every week and the second pays $3.00 an hour with a 5¢ raise every week. Assuming a 40-hr work week, after how many weeks do they pay the same?

If we let w denote the number of weeks, y the rate on the first job, and z the rate on the second job, we have

$$y = 2.75 + .1w$$
$$z = 3.00 + .05w$$

The question is, for what value of w is $y = z$? If $y = z$, then

$$2.75 + .1w = 3.00 + .05w$$
$$.05w = .25$$
$$5w = 25$$
$$w = 5$$

Thus at the end of 5 weeks, the two jobs pay the same. Notice that the fact of a 40-hr work week was redundant, that is, it does not enter into the problem. (Can you decide which job is better for a 12-week summer vacation?) □

When we translate the real problem situation into the world of mathematics, solve our mathematical problem, and then translate back, sometimes not all of our mathematical answers make sense in the practical situation.

Example 6

A rectangular flower garden is 5 times as long as it is wide. If the total area is 720 sq ft, what are the dimensions of the garden?

We know $A = lw$ where A denotes area, l denotes length, and w denotes width. We are given that $A = 720$ and $l = 5w$. Thus we have the equation

$$720 = (5w)w$$
$$720 = 5(ww)$$
$$144 = w \cdot w$$
$$w = 12 \quad \text{or} \quad w = -12 \quad \text{and} \quad l = 60 \quad \text{or} \quad l = -60$$

Since negative dimensions for a flower garden make no sense, the width is 12 and the length is 60. □

The problems in the next exercise set can all be solved by the techniques studied in Sections 1–6. Take sufficient time to make sure you understand the problem and analyze the relationships among the quantities under consideration. Treat each problem individually, do not try to force one problem into the form of another. The first few problems are mathematical puzzles but the others are problems of the sort that arise in the work-a-day world. We include the puzzles because they are a little easier. Some of the problems rely on some basic formulas or relationships that you may not know or may have forgotten. Such information is available to you; look it up.

EXERCISES

1. The sum of two numbers is 14.3 and their difference is 4.7. What are the numbers?

2. Suppose you have three numbers such that their sum is 24, the sum of the first two is 7 and the sum of the last two is 13. What are the numbers?

3. Suppose $y = mx + b$ for all numbers x. If $y = 3$ when $x = 1$ and $y = 7$ when $x = -1$, what are m and b?

4. Suppose you have an average of 72.4 before the final exam in a course. If the final exam counts $\frac{1}{3}$ of the grade, what must you receive on the final in order to have at least a 70 average for the course?

5. The average, A, of a student is given by the formula $A = P/C$ where P indicates the quality points earned and C is the number of credits earned. The number of quality points earned in a course is the product of the number of credits in a course and the number which corresponds to the grade. At most universities A corresponds to 4, B to 3, C to 2, D to 1, and F to 0. The total number of quality points P is determined by summing those earned for each course. Suppose you have an average of 2.36, have earned 45 credits, and are taking 12 credits this term. What must your average be for the term in order to bring your average up to a 2.40?

6. Suppose 180 credits are required for graduation; you have earned a total of 165 credits with an average of 2.44. If you are taking 15 credits this term and an average of 2.25 is required for graduation, what is the minimal average you can earn this term and still graduate?

7. Suppose you have two outboard motors, the older of which burns a mixture of 1 part oil to 20 parts gasoline and the newer of which uses 1 part oil to 50 parts of gasoline. If you have an abundant supply of gasoline for the newer motor, how much oil would you have to add to 10 gal of 1 : 50 mixture to make it suitable to burn in the old motor?

8. Suppose you have your choice of two fertilizers, type A and type B. To fertilize your fields with type A cost $4 more per acre than with type B but type A brings a 5 percent higher yield than does type B. What must your gross receipts per acre be in order for type A to be a more profitable investment than type B?

9. A nut shop manager wishes to mix peanuts, cashews, almonds, and pecans in order to sell a 10-lb mixture of nuts. If peanuts are 49¢ a pound, cashews $1.19 a pound, almonds $1.09 a pound, and pecans $1.24 a pound, and if the amounts of cashews, almonds, and pecans are the same, how many pounds of each kind of nut should he use to produce a ten-pound mixture selling at $.99 a pound?

10. If a truck leaves a turnpike service station at 10:15 traveling eastward at 60 mph and a car leaves the same station at 10:20 and travels 70 mph, at what time will the car catch the truck?

11. A tank contains 12 gal of 5 percent salt brine. How much water must be evaporated in order to yield a solution of 7 percent salt?

12. Suppose you have two numbers such that the absolute value of their sum is 10 and the absolute value of their difference is 2. What could the numbers be?

13. Suppose a baseball player has a batting average of .240 after 310 times at bat. If he can expect to bat about 200 more times during the season, how many hits must he get in order to assure that he bats at least .275?

14. One orthodontist makes no initial charge but charges $50 a month for the period of time that the patient wears the braces. Another charges $250 at the time the braces are installed and $25 a month for the period of time that the

patient wears the braces. How many months would a patient have to wear the braces in order for the charges to be less from the second orthodontist?

15. The air fare from Dallas is $156 but the special excursion package price of $104 each is available if at least 10 people travel together. It is not really necessary for 10 people to go; it is only necessary for 10 fares to be paid at the excursion rate and for those going to travel at the same time on the same plane. How many people actually need to go in order to save money by buying the 10 tickets?

16. As with everything else, the price of paper keeps increasing. Suppose that you are a newspaper publisher and you find that by changing type face and the printing style of your newspaper, you can cut the amount of paper used by 7 percent per week. What kind of percentage increase in paper price can you offset by making these changes?

17. Suppose you are in a federal income tax bracket in which the government takes 25 percent of the first $200 of any annual increase in your income and the state tax takes 3 percent of any increase in your income. How much of a raise do you need in order to increase your monthly net income by $50?

18. Automotive engineers have determined that current automobile engines can easily be adapted to operate on a mixture of 80 percent gasoline and 20 percent methol alcohol. If the cost of gasoline delivered to the station is 39¢ per gallon, the retail price of gasoline is 54¢ per gallon, and the cost of producing and delivering the methol alcohol to the station is 15¢ per gallon, what would be the price of the 80-20 mixture given the same percentage of mark up?

19. A sportscaster suggests that the penalty for a false start in the 440-yd run should be a 5-yd set back, that is, the false starter must give the others a 5-yd head start. Suppose you are penalized and still win with a time of 57 sec. What was your average time for the actual 440? (Does this seem like a reasonable penalty?)

20. Suppose two companies sell batteries for your car, one with a 5-year prorated guarantee and one with a 4-year prorated guarantee. Suppose also that the battery with the longer guarantee sells for a firm price of $30 and you are dickering with the dealer over the price of the one with the shorter guarantee. Assuming that nothing lasts beyond its guarantee, what would the price have to be in order for the battery with the 4-year guarantee to be the better buy for you?

21. Suppose you hire two typists to type a paper for you. One types 45 words per minute and the other 70 words per minute. If the paper contains about 15,000 words, how long will it take them to type the paper assuming that they work the same amount of time?

22. Refer to Exercise 21. Suppose you pay the slower of your typists $2 per hour. What should you pay the second if you wish to pay according to the amount of work done? What will the typing cost you at these prices?

23. The perimeter of a rectangle is 92 ft. What are the dimensions if the length is 5 times the width?

24. The circumference of one circle is 7 times that of another. If the smaller circle has a radius of 22, what is the radius of the larger?

25. In a particular city, the labor department finds an overall unemployment rate of 5.2 percent with an unemployment rate of 3.1 percent among whites. If non-

whites make up 27 percent of the population in the city, what is the unemployment rate among these minorities?

26. Refer to Exercise 25. If the population of the city is 520,000, how many nonwhites would have to find employment in order for the rates to be the same?

27. Suppose you and a friend are canoeing on a stream and you wish to determine your speed in still water. If you know that a particular bridge is just $\frac{1}{2}$ mi from your starting point and you paddle downstream to the bridge in 4.2 min but it takes you 7.8 min to return, what is your speed in still water? What is the rate of the current?

28. Suppose Mr. Bunker wishes to make some green paint by mixing yellow and blue together. His perfect memory and judgment tell him that he should use 5 parts of yellow to 2 parts of blue and so he mixes a gallon this way. To his amazement he discovers that it is too light. Amid gales of laughter (from others) he leaves for the corner bar and turns the job over to his son-in-law (the one leading the laughter). The son-in-law relies on his equally perfect memory and judgment and, with his usual disdain for money, starts with new paint and mixes a gallon using 7 parts yellow and 4 parts blue. He is amazed to find that his mixture is too dark and sheepishly leaves to join his father-in-law at the saloon, leaving the job to his wife and mother-in-law. The mother-in-law, by first mixing very small portions, determines that the proper mixture is 2 parts yellow and 1 part blue. Instead of starting again, however, she wishes to use the light and dark mixtures to make one that is just right. How much of each should she use to obtain a gallon of the right shade?

29. By testing your car at 55 mph and 70 mph under similar driving conditions, you determine that it yields 18.5 mpg at 55 mph and 14 mpg at 70 mph.
    (a) If gasoline is 54¢ per gallon, how much money do you save on a 400-mi trip by driving 55 mph instead of 70 mph?
    (b) How much longer does the trip take at 55 mph?
    (c) How much per hour are you earning? (Your savings are your earnings.)
    (d) What would the price of gasoline have to be in order for you to earn at least $2 per hour?
    *(e) Does the 400-mi distance enter into your answers for parts (c) and (d)? Explain.

30. An aluminum siding contractor assures you that if he puts siding on your house, it will not need any painting or maintenance for at least 20 years. Suppose that previous experience tells you that in order to keep the house in good condition, it must be painted at least every 4 years. Furthermore, assume that the house needs painting now and that it will cost $850 to do the job.
    (a) If the siding job costs $4500, which will cost you more over the long run, painting or siding?
    (b) Perhaps the question in part (a) was too simple. If you paint the house now and invest the balance of $3650 (the difference between painting and siding) in a savings account at 5 percent per year and use the interest in painting the house, then which costs more? (Ignore the compounding of interest.)
    (c) What would the costs of painting have to be for the costs of both to be the same over a 20-year period?

31. Mr. Pinchpenny operates a hamburger stand and wishes to spend as little as

possible on ketchup, mustard, and so on. It is his practice to add water to the ketchup to make it go farther and he has found that if he adds one pint of water to each gallon of ketchup, no one can tell the difference. On April 16 he feels that he needs to save even more money and decides to add a quart of water to each gallon of ketchup and mixes up 5 gal of his new mixture using 1 gal of water and 4 gal of ketchup. Usually a 5-gal supply would last him for a week but he notices that 2 gal of the new mixture are used in only one day with no appreciable change in the number of sandwiches sold. The mixture is so thin that the customers are pouring more than they expect to. He realizes that he should go back to his original mixture. How much pure ketchup should he add to the remaining 3 gal?

32. A publisher of a monthly magazine has determined, from experience, that the following relationships hold:

  (i) The number S of subscription customers is related to A, the annual subscription price (in dollars) by the equation $S = -30,000A + 225,000$.
  (ii) The monthly costs C of printing and mailing the magazine and operating the business are given by the equation $C = .13N + .03S + 20,000$, where N is the number of magazines printed per month.
  (iii) The advertising revenue is given by $R = .11N$.
  (iv) The newsstand sales income T is given by $T = pn$ where p is the net price and n is the number sold on newsstands.

Let us assume for this problem that $n = 10,000$ and $N = S + 10,000$, that is, newsstand sales are 10,000 and that they sell all they print. Assume also that $p = 35\cent$.
  (a) If the annual subscription price is $3.60, what is the net profit per month?
  (b) If postage rates are changed, the increased costs must be passed on to the subscription customers. (The $.03S$ term in the equation for C is the present mailing cost. If mailing costs change, this changes.) Determine the net profit per month if the cost of mailing is increased to $8\cent$ per copy and this is passed on to the subscribers.
  (c) What would be the net profit if mailing costs remain the same and the subscription cost is lowered to $3 per year?

33. Suppose that the number N of men enlisting in the army each month from Gotham City is given by the equation $N = Kr + C$ where r is the unemployment rate in the city and K and C are constants.
  (a) Determine K and C if 1200 men enlisted when the rate was 4 percent and 1356 enlisted when the rate was 5.2 percent.
  (b) How many would you expect to enlist if the rate goes up to 6 percent?

34. The monthly rent in a particular store is given by the equation $R = mS + b$ where R denotes the rent and S denotes the monthly sales.
  (a) Determine m and b if the rent is $500 with sales of $15,000 and it is $550 with sales of $17,000.
  (b) What would you expect the rent to be if the sales are $21,000?

3

The Arithmetic of Algebraic Expressions

An algebraic expression is any combination of letters, numerals, and operational symbols which, when numerals are substituted for letters, yields a name for a number.

Example 1

$$\frac{x + 3}{x - 2} \quad \square$$

Example 2

$$xy - uv + zw + 5p \quad \square$$

Example 3

$$\pi d \quad \square$$

Example 4

$$\tfrac{1}{2}bh \quad \square$$

We limit the substitutions to those numerals which yield a meaningful form.

79

Example 5

We consider

$$\frac{x + 7}{x - 3}$$

to be an algebraic expression but do not allow 3 to be substituted for x because this would yield a meaningless form $10/0$. ☐

Since algebraic expressions represent numbers and since numbers can be added, subtracted, multiplied, and divided we can also perform these arithmetic operations on algebraic expressions. In order to solve problems more complex than those treated in Chapter 2, you need skill in performing these more complicated computations. The rules governing the arithmetic of algebraic expressions are just those that govern the arithmetic of numbers. There are no differences between the principles of elementary arithmetic and the principles of the arithmetic of algebra. Since algebraic expressions tend to be more complicated than most numerical ones, the techniques are more complex. *If you know what the symbols mean and know what principles are being applied, you will avoid most difficulties.*

1. The Arithmetic of Polynomials

In developing the arithmetic of algebraic expressions, it is usually best to begin with the simplest forms and build up to more complicated expressions. The simplest expressions, of course, are those involving just one numeral or variable and no operational symbol, that is, expressions such as "93" or "x." We have dealt with such expressions as well as forms such as "$2x + 5$" and "$7x - 3y$" in Chapter 2. In this section we treat the arithmetic of a slightly more complicated class of expressions, those known as *polynomials*.

The first step is to develop some notation that can be used to simplify expressions and eliminate a great deal of writing.

If we wish to write the product of seven 5's we could write

$$5 \cdot 5 \cdot 5 \cdot 5 \cdot 5 \cdot 5 \cdot 5$$

It is more convenient, however, to write 5^7. The "7" in this expression is called an *exponent*. Similarly, $(-4)(-4)(-4)(-4)(-4)$ can be written $(-4)^5$. In general:

If a is a real number and n is a positive integer, then $a^n = a \cdot a \ldots a$ where a appears n times. The number n is called an *exponent*.

From the definition we see that $a^1 = a$ and $a^2 = a \cdot a$. The expression a^n is read "a to the nth power." The expression a^2 is usually read "a squared" or "the square of a" and a^3 is read "a cubed" or "the cube of a." These special names stem from geometry in that the area of a square with side s is s^2 and the volume of a cube with side s is s^3. To compute a^n or just write a^n is said to "raise a to the nth power."

Consider now the following products:

$$(5^2)(5^4) = (5 \cdot 5)(5 \cdot 5 \cdot 5 \cdot 5) = 5^6$$
$$3^5 \cdot 3^4 = (3 \cdot 3 \cdot 3 \cdot 3 \cdot 3)(3 \cdot 3 \cdot 3 \cdot 3) = 3^9$$
$$7^2 \cdot 7^3 = (7 \cdot 7)(7 \cdot 7 \cdot 7) = 7^5$$

In general:

If a is a real number with m and n positive integers, then

$$a^m \cdot a^n = \underbrace{(a \cdot a \cdot \ldots \cdot a)}_{m \text{ times}} \cdot \underbrace{(a \cdot a \cdot \ldots \cdot a)}_{n \text{ times}} = a^{m+n}$$

that is,

$$a^m \cdot a^n = a^{m+n}$$

This is the first law of exponents.

Example 6

$$2^4 \cdot 2^5 = 2^9 \ \square$$

Example 7

$$7^3 \cdot 7^2 = 7^5 \ \square$$

Example 8

$$(19)^{14} \cdot (19)^7 = (19)^{21} \ \square$$

Notice that this law depends only on the associative law and the *meaning of exponents*.

If we raise a power of a number to a power, we obtain a power of the original number.

$$(3^2)^4 = 3^2 \cdot 3^2 \cdot 3^2 \cdot 3^2 = 3^{2+2+2+2} = 3^8$$
$$(7^3)^5 = 7^3 \cdot 7^3 \cdot 7^3 \cdot 7^3 \cdot 7^3 = 7^{3+3+3+3+3} = 7^{15}$$

In general:

If a is a real number with m and n positive integers, then

$$(a^m)^n = \underbrace{a^m \cdot a^m \cdot \ldots \cdot a^m}_{n \text{ times}} = a^{m+m \ldots + m} = a^{mn}$$

that is,

$$(a^m)^n = a^{mn}$$

This is the second law of exponents.

Example 9

$(4^7)^3 = 4^{21}$ ☐

Example 10

$[(-2)^5]^3 = (-2)^{15}$ ☐

Example 11

$[(.01)^4]^7 = (.01)^{28}$ ☐

Notice that this law depends on the associative law, the first law of exponents, and the *meaning of exponents*.

Since the product of two positive numbers is positive, the product of any number of positive numbers is positive, and in particular, if $a > 0$, then $a^n > 0$ for all positive integers n. The product of two negative numbers is positive and, in fact, the product of any even number of negative numbers will be positive since we can first compute the product of pairs and then multiply the corresponding positive numbers. Thus if $a < 0$ and n is an even positive integer, $a^n > 0$. On the other hand, the product of an odd number of negative numbers must be negative and, in particular, if $a < 0$ and n is odd, then $a^n < 0$.

Powers of products can also be simplified and, in fact, are products of powers:

$$(4 \cdot 7)^2 = (4 \cdot 7)(4 \cdot 7) = (4 \cdot 4)(7 \cdot 7) = 4^2 \cdot 7^2$$

$$(2 \cdot 3)^4 = (2 \cdot 3)(2 \cdot 3)(2 \cdot 3)(2 \cdot 3) = (2 \cdot 2 \cdot 2 \cdot 2)(3 \cdot 3 \cdot 3 \cdot 3) = 2^4 \cdot 3^4$$

$$(5 \cdot 8)^3 = (5 \cdot 8)(5 \cdot 8)(5 \cdot 8) = (5 \cdot 5 \cdot 5)(8 \cdot 8 \cdot 8) = 5^3 \cdot 8^3$$

In general:

If x and y are real numbers and n is a positive integer, then

$$(xy)^n = (xy)(xy)(xy) \ldots (xy) = (x \cdot x \cdot \ldots \cdot x)(y \cdot y \cdot \ldots \cdot y) = x^n \cdot y^n$$

We have the third law of exponents:

$$(xy)^n = x^n y^n$$

Example 12

$(3 \cdot 7)^4 = 3^4 \cdot 7^4$ ☐

Example 13

$(5 \cdot 6)^7 = 5^7 \cdot 6^7$ ☐

Example 14

$$(2 \cdot 3)^{10} = 2^{10} \cdot 3^{10} \ \square$$

This law depends on the associative and commutative laws and the *meaning of exponents.* Powers of quotients or fractions follow a similar law.

$$\left(\frac{3}{5}\right)^2 = \frac{3}{5} \cdot \frac{3}{5} = \frac{3 \cdot 3}{5 \cdot 5} = \frac{3^2}{5^2}$$

$$\left(\frac{2}{3}\right)^4 = \frac{2}{3} \cdot \frac{2}{3} \cdot \frac{2}{3} \cdot \frac{2}{3} = \frac{2 \cdot 2 \cdot 2 \cdot 2}{3 \cdot 3 \cdot 3 \cdot 3} = \frac{2^4}{3^4}$$

$$\left(\frac{4}{7}\right)^3 = \frac{4}{7} \cdot \frac{4}{7} \cdot \frac{4}{7} = \frac{4 \cdot 4 \cdot 4}{7 \cdot 7 \cdot 7} = \frac{4^3}{7^3}$$

In general:

If x and y are real numbers with $y \neq 0$ and n is a positive integer, then

$$\left(\frac{x}{y}\right)^n = \frac{x}{y} \cdot \frac{x}{y} \cdot \ldots \cdot \frac{x}{y} = \frac{x \cdot x \cdot \ldots \cdot x}{y \cdot y \cdot \ldots \cdot y} = \frac{x^n}{y^n}$$

that is,

$$\left(\frac{x}{y}\right)^n = \frac{x^n}{y^n}$$

This is the fourth law of exponents and depends on the meaning of exponents and the principles which govern the multiplication of fractions.

Example 15

$$\left(\frac{2}{3}\right)^2 = \frac{2^2}{3^2} = \frac{4}{9} \ \square$$

Example 16

$$\left(\frac{4}{7}\right)^3 = \frac{4^3}{7^3} \ \square$$

Example 17

$$\left(\frac{1}{4}\right)^5 = \frac{1^5}{4^5} = \frac{1}{4^5} \ \square$$

In the above discussions we established the four basic laws of exponents which are stated together below for the convenience of the student.

If a and b are real numbers and m and n are positive integers, then

(i) $a^m \cdot a^n = a^{m+n}$
(ii) $(a^m)^n = a^{mn}$
(iii) $(ab)^n = a^n b^n$
(iv) If $b \neq 0$,

$$\left(\frac{a}{b}\right)^n = \frac{a^n}{b^n}$$

An algebraic expression that consists of only one constant or variable or that is an indicated product of several constants or variables is called a *monomial*. Notice that monomials do not involve addition, subtraction, or division. An expression which is the sum of two monomials is called a *binomial*; an expression which is the sum of three monomials is called a *trinomial*, and any expression consisting of one monomial or the sum of several monomials is called a *polynomial*. The monomials that are added together are called the *terms* of the polynomial and the numerical parts of the terms are called *coefficients*. Expressions such as $4x^2 - 3xy - 7y^2$ are also called polynomials because subtraction can always be rewritten using addition.

Example 18

$$4x^2 - 3xy - 7y^2 = 4x^2 + (-3)xy + (-7)y^2 \quad \square$$

A polynomial in which the variable parts of the monomials are the same can be simplified by using the distributive laws and completing the indicated computation with the constant factors.

Example 19

$$5xy^2 + 7xy^2 = (5 + 7)xy^2 = 12xy^2 \quad \square$$

Example 20

$$3xyz - 7xyz = (3 - 7)xyz = -4xyz \quad \square$$

Example 21

$$8x^2 - 3x^2 + 4x^2 = (8 - 3 + 4)x^2 = 9x^2 \quad \square$$

In practice we do not need to write down the intermediate steps but can do the computation mentally. The process of simplifying such polynomials is sometimes called "collecting like terms," *like terms* being monomials with the same variable parts.

If we wish to add or subtract two polynomials, frequently the simplest way is to write one polynomial below the other with like terms in the same columns and then add or subtract in each column. The commutative and associative laws enable us to use this procedure.

Example 22

Add

$$4x^3 - 7x^2y - 2xy^2 + 5y^3$$
$$5x^3 + 2x^2y + 9xy^2 - 3y^3$$
$$\overline{9x^3 - 5x^2y + 7xy^2 + 2y^3} \ \square$$

Example 23

Subtract

$$7x^4 + 2x^3 - \ \ 9x^2 + 18x - \ \ 7$$
$$12x^4 - 3x^3 + \ \ 5x^2 + \ \ 2x + \ \ 3$$
$$\overline{-5x^4 + 5x^3 - 14x^2 + 16x - 10} \ \square$$

Three or more polynomials may be added in the "columnar" method.

Example 24

Add

$$5x^4 - 6x^3 + 7x^2 - 23x + 17$$
$$-6x^4 - 2x^3 - 9x^2 + \ \ 4x + 11$$
$$-4x^4 + 3x^3 + 5x^2 + 11x - \ \ 5$$
$$\overline{-5x^4 - 5x^3 + 3x^2 - \ \ 8x + 23} \ \square$$

Example 25

To add $5x^4 + 3x^2 - 7x + 3$ and $7x^4 + 6x^3 + 5x^2 + 8$ we just note that the first polynomial could be rewritten with 0 as the coefficient of the term involving x^3 and the second could be rewritten with 0 as the coefficient of the term involving x.

$$5x^4 + 0x^3 + 3x^2 - 7x + \ \ 3$$
$$7x^4 + 6x^3 + 5x^2 + 0x + \ \ 8$$
$$\overline{12x^4 + 6x^3 + 8x^2 - 7x + 11}$$

It is more common not to write the terms with zero coefficient but just leave a space in the corresponding column.

$$\begin{array}{l} 5x^4 \qquad\quad + 3x^2 - 7x + 3 \\ 7x^4 + 6x^3 + 5x^2 \qquad\quad + 8 \\ \hline 12x^4 + 6x^3 + 8x^2 - 7x + 11 \ \square \end{array}$$

Multiplying polynomials can be accomplished by applying the distributive laws several times and combining like terms.

Example 26

$$(x^2 + 2x + 3)(x^3 - 2x^2 + 5x - 7)$$
$$= x^2(x^3 - 2x^2 + 5x - 7) + 2x(x^3 - 2x^2 + 5x - 7)$$
$$\quad + 3(x^3 - 2x^2 + 5x - 7)$$
$$= x^5 - 2x^4 + 5x^3 - 7x^2 + 2x^4 - 4x^3 + 10x^2$$
$$\quad - 14x + 3x^3 - 6x^2 + 15x - 21$$
$$= x^5 + 4x^3 - 3x^2 + x - 21 \ \square$$

These same laws can be applied as they are when we have numerals representing numbers, that is, write one below the other, multiply the first polynomial by each term in the second, write these products in columnar form, and add. It is customary to start the multiplication on the left although it is not necessary.

Example 27

$$\begin{array}{l} 2x^2 + 3xy - 5y^2 \\ 5x^2 - 7xy - 4y^2 \\ \hline 10x^4 + 15x^3y - 25x^2y^2 \\ \quad - 14x^3y - 21x^2y^2 + 35xy^3 \\ \qquad\quad - 8x^2y^2 - 12xy^3 + 20y^4 \\ \hline 10x^4 + \quad x^3y - 54x^2y^2 + 23xy^3 + 20y^4 \ \square \end{array}$$

Example 28

$$\begin{array}{l} x^3 - 2x^2 + 5x - 7 \\ \qquad\quad x^2 + 2x + 3 \\ \hline x^5 - 2x^4 + 5x^3 - 7x^2 \\ \quad 2x^4 - 4x^3 + 10x^2 - 14x \\ \qquad\quad 3x^3 - 6x^2 + 15x - 21 \\ \hline x^5 \qquad\quad + 4x^3 - 3x^2 + x - 21 \ \square \end{array}$$

Division of one polynomial by another can be accomplished by a method similar to that used for long division of numbers. If an exact quotient exists, the method will produce it but if not, we obtain a quotient and a remainder as in arithmetic. The remainder will not be "smaller" than the divisor but if we are computing with polynomials containing only one variable, say x, then the highest power of x appearing in the remainder is lower than the highest power appearing in the divisor.

Example 29

$$
\begin{array}{r}
x^3 + 8x^2 + 17x + 53 \\
x - 3 \overline{)x^4 + 5x^3 - 7x^2 + 2x - 5} \\
\underline{x^4 - 3x^3} \\
8x^3 - 7x^2 \\
\underline{8x^3 - 24x^2} \\
17x^2 + 2x \\
\underline{17x^2 - 51x} \\
53x - 5 \\
\underline{53x - 159} \\
154
\end{array}
$$

The quotient is $x^3 + 8x^2 + 17x + 53$ and the remainder is 154. $\square$

Example 30

$$
\begin{array}{r}
5x^3 + 2x^2 + (17/2)x + 35/2 \\
2x^2 - 2x - 1 \overline{)10x^5 - 6x^4 + 8x^3 + 16x^2 - 20x + 17} \\
\underline{10x^5 - 10x^4 - 5x^3} \\
4x^4 + 13x^3 + 16x^2 \\
\underline{4x^4 - 4x^3 - 2x^2} \\
17x^3 + 18x^2 - 20x \\
\underline{17x^3 - 17x^2 - (17/2)x} \\
35x^2 - (23/2)x + 17 \\
\underline{35x^2 - 35x - 35/2} \\
(47/2)x + 69/2
\end{array}
$$

The first power is the highest power of x that can appear in the remainder because the second power of x is the highest power of x appearing in the divisor. $\square$

Sometimes some terms will not appear in the dividend but will appear in the divisor or arise in the computational process. Such polynomials can be rewritten using some zero coefficients for these terms just as we did for addition. We then proceed as before.

Example 31

$$
x - 2 \overline{)x^4 - 3}
$$

We can rewrite and compute as before.

$$
\begin{array}{r}
x^3 + 2x^2 + 4x + 8 \\
x - 2\overline{)x^4 + 0x^3 + 0x^2 + 0x - 3} \\
\underline{x^4 - 2x^3} \\
2x^3 + 0x^2 \\
\underline{2x^3 - 4x^2} \\
4x^2 + 0x \\
\underline{4x^2 - 8x} \\
8x - 3 \\
\underline{8x - 16} \\
13
\end{array}
$$

The quotient is $x^3 + 2x^2 + 4x + 8$ and the remainder is 13. $\square$

Once you get some practice you will probably eliminate the rewriting and just leave space at the appropriate points.

Example 32

$$
\begin{array}{r}
x^3 - x^2 + 2x \\
x^2 + x + 2\overline{)x^5 + 3x^3 - 8} \\
\underline{x^5 + x^4 + 2x^3} \\
- x^4 + x^3 \\
\underline{- x^4 - x^3 - 2x^2} \\
2x^3 + 2x^2 \\
\underline{2x^3 + 2x^2 + 4x} \\
- 4x - 8
\end{array}
$$

The quotient is $x^3 - x^2 + 2x$ and the remainder is $-4x - 8$. $\square$

EXERCISES

1. Write the following without using exponents.
   (a) $2^2 \cdot 2^3$ (b) $(\frac{2}{3})^3$ (c) $(\frac{1}{4})^2$ (d) $(7 \cdot 3)^2$
   (e) $(\frac{7}{3})^2$ (f) $(.1)^5$ (g) $(.01)^3$ (h) $(\frac{4}{9})^2(9.4)^3$

2. Evaluate the following expressions if $a = 2$, $b = -3$, $x = 4$, and $y = -5$.
   (a) $a^2 - 3ab + 2b^2$
   (b) $a^2x^2 + 4ax^3 + 5a^2x - 2a^3 + 4x^2$
   (c) $3x^2 - 7xy + 5y^2$
   (d) $6x^2 - 7xb + b^2$
   (e) $-4x^2a + 5xa^2 - 7xa + 3a - 10x$
   (f) $x^2y^2 + 3xy^2 - 7x^2y - 4y^2$

(g) $y^3b^2 + 3y^2b^3 + 5yb - 7y^2b^2$

(h) $x^2y^3ab^2 - 7xya^2b + 13x^2yab^3$

3. Compute the following products.

(a) $(x^4 - 7x^3 + 3x^2 - 19x + 11)(2x^2 - 5x + 7)$

(b) $(3x^3 - 5x^2 + 17x - 9)(4x^3 + 9x^2 - 3x + 2)$

(c) $(7x^5 - 6x^3 + 5x - 9)(6x^5 + 11x^4 - 17x + 5)$

(d) $(x^2y^3 + 3x^3y^2 - 5x^2y^2 + 9xy^2 + 3x^2y - 9xy)(2x^2y^2 - 5xy^2 - 7x^2y)$

(e) $(x^2y^3z^2 + 3xy^4z^2 + 5x^2yz)(5xy^2z + 3x^2yz^2 + 2x^2y^2z^2)$

(f) $(2x^4 - 7x^3 + 8x^2 - 5x + 7)(3x^2 - 10x + 6)$

(g) $(5x^5 - 6x^4 - 3x^3 + 9x^2 + 3x - 17)(4x^3 - 7x^2 + 13x - 5)$

(h) $(\frac{2}{3}x^3 - \frac{1}{6}x^2 + \frac{2}{5}x - \frac{1}{4})(\frac{1}{2}x^2 - \frac{1}{3}x - \frac{1}{5})$

(i) $(.6x^4 - 5.2x^3 + .48x^2 + 7.2x + 3.6)(4.2x^3 - .7x^2 + 3.4x - 7.1)$

(j) $(36x^3 - 71x^2 - 54x + 92)(23x^2 + 41x + 19)$

4. Compute the following products.

(a) $(x + y)(x + y)$ (b) $(a - b)(a - b)$

(c) $(u + v)(u - v)$ (d) $(a - b)(a^2 + ab + b^2)$

(e) $(a + b)(a^2 - ab + b^2)$ (f) $(x - y)(x^3 + x^2y + xy^2 + y^3)$

(g) $(a - b)(a^4 + a^3b + a^2b^2 + ab^3 + b^4)$

(h) $(a + b)(a^4 - a^3b + a^2b^2 - ab^3 + b^4)$

(i) $(a + b)^3$ (j) $(a + b)^4$

5. Perform the indicated divisions.

(a) $(x^4 + 5x^3 - 7x^2 + 3x - 2) \div (x - 2)$

(b) $(x^3 - 7x^2 + 3x + 17) \div (x + 1)$

(c) $(x^5 - 6x^3 + 5x^2 + 3x - 2) \div (x^2 + x - 2)$

(d) $(6x^7 - 3x^6 + 5x^5 - 7x^4 + 2x^3 + 6x^2 + 22x + 19) \div (2x^2 - 3x + 1)$

(e) $(3x^5 - 7x^4 + 5x^3 - 2x^2 + 3x - 5) \div (x^2 - x + 2)$

(f) $(.6x^4 + 4.3x^3 - .7x^2 + 3.4x + 1.5) \div (x^2 + .2x - .3)$

(g) $(4x^3 - 7x^2 + 9x - 6) \div (x - 1)$

(h) $(10x^4 - 20x^3 + 40x^2 + 16x + 12) \div (2x^2 + 4x - 6)$

(i) $(2x^5 + 2x^4 + 4x^3 - 6x^2 + 10x - 5) \div (2x^2 + x - 3)$

(j) $(x^3 - 1) \div (x - 1)$ (k) $(x^5 + 1) \div (x + 1)$

(l) $(x^4 - 16) \div (x - 2)$ (m) $(x^7 - 1) \div (x - 1)$

(n) $(x^3 - y^3) \div (x - y)$ (o) $(x^4 - y^4) \div (x - y)$

(p) $(a^5 - b^5) \div (a - b)$ (q) $(a^6 - b^6) \div (a - b)$

6. Compute the following products.

(a) $(x + y)(x - 2y)$ (b) $(2a - b)(a + 3b)$

(c) $(6y + z)(5y - 7z)$ (d) $(5x - 2y)(3x + 7y)$

(e) $(7a - 3b)(5a - 2b)$ (f) $(8x + 3y)(4x + 5y)$

(g) $(10x + 7y)(3x + 5y)$ (h) $(x - 8y)(2x - 5y)$

(i) $(3x - 13y)(5x - 3y)$ (j) $(9x + 5y)(11x - 7y)$

2. The Product of Two Binomials

If we wish to multiply two binomials, it is frequently more convenient to write them side by side and compute term-by-term rather than writing the product in column form.

Example 1

$$(a + 2b)(2c - 3d) = a(2c - 3d) + 2b(2c - 3d)$$
$$= 2ac - 3ad + 4bc - 6bd \ \square$$

Example 2

$$(2x + 3)(x - 5) = 2x(x - 5) + 3(x - 5)$$
$$= 2x^2 - 10x + 3x - 15$$
$$= 2x^2 - 7x - 15 \ \square$$

Example 3

$$(a + 2b)(3a + b) = a(3a + b) + 2b(3a + b)$$
$$= 3a^2 + ab + 6ab + 2b^2$$
$$= 3a^2 + 7ab + 2b^2 \ \square$$

In the last two examples, the variables involved were the same in both binomials and the result is a trinomial. There are four multiplications involved in such computations and the pattern is illustrated by the diagrams in the following examples.

Example 4

$$(2x + 5y)(3x + 7y) = 6x^2 + 14xy + 15xy + 35y^2$$
$$= 6x^2 + 29xy + 35y^2 \ \square$$

Example 5

$$(5a - 7b)(2a + 11b) = 10a^2 + 55ab - 14ab - 77b^2$$
$$= 10a^2 + 41ab - 77b^2 \ \square$$

Example 6

$$(x - 5)(3x - 7) = 3x^2 - 7x - 15x + 35$$
$$= 3x^2 - 22x + 35 \ \square$$

The general procedure is:

(i) Compute the product of the first·terms in the two binomials.
(ii) Compute the product of the first term in the first binomial with the second term in the second binomial and the product of the second term in the first binomial with the first term in the second binomial. If they are like terms, combine them.
(iii) Compute the product of the last terms in the two binomials.
(iv) Write the sum of all three products.

After some practice, you should be able to do step (ii) mentally.

Example 7

$$(4x - 7y)(5x + 2y) = 20x^2 - 27xy - 14y^2 \;\square$$

Example 8

$$(3y - 5)(2y - 11) = 6y^2 - 43y + 55 \;\square$$

Example 9

$$(6x + 5)(4x + 3) = 24x^2 + 38x + 15 \;\square$$

EXERCISES

Compute the following products.

1. $(5x + 1)(2x + 3)$
2. $(6x - 5y)(x - 2y)$
3. $(3a + b)(7a + 2b)$
4. $(10x - 3y)(2x - y)$
5. $(12x + 5y)(2x - 3y)$
6. $(16x - 1)(8x - 3)$
7. $(4x + 5z)(7x - 2z)$
8. $(x + 4y)(2x + 7y)$
9. $(5x + 3)(2x + 1)$
10. $(9x + 7y)(5x - 3y)$
11. $(y + 10z)(2y + 3z)$
12. $(x - 7y)(x - 5y)$
13. $(4x + 5y)(3x + y)$
14. $(7a - b)(5a - b)$
15. $(12a + 17b)(3a - 7b)$
16. $(6a + 5b)(a - 11b)$
17. $(3x - 2)(5x - 2)$
18. $(13x + 2y)(11x - 5y)$
19. $(2x + 11y)(3x + 13y)$
20. $(8x - 7z)(10x + 3z)$
21. $(x + 1)(x - 1)$
22. $(a + 2b)(a - 2b)$
23. $(3x - y)(3x + y)$
24. $(x + 5y)(x - 5y)$
25. $(2x + 3y)(2x - 3y)$
26. $(5x - 2y)(5x + 2y)$
27. $(x + 2)(x - 2)$
28. $(4x + 3)(4x - 3)$
29. $(6x - 7)(6x + 7)$
30. $(7x - 2z)(7x + 2z)$
31. $(x + 2y)(x + 2y)$
32. $(3x - y)(3x - y)$
33. $(a + 2b)(a + 2b)$
34. $(5x - y)(5x - y)$
35. $(7a + 2b)(7a + 2b)$
36. $(5a - c)(5a - c)$
37. $(10x + 3y)(10x + 3y)$
38. $(8x - 7)(8x - 7)$
39. $(9x + 2)(9x + 2)$
40. $(13x + 14y)(13x + 14y)$

3. Factoring Polynomials

Multiplication of algebraic expressions is governed by the associative, commutative, and distributive laws and we have seen how indicated products can be simplified by using these laws. Many times, however, an expression is more useful if it is written as a product of simpler expressions rather than as a polynomial with several terms. In this section we will be concerned with rewriting polynomials in this (sometimes) more useful form.

Example 1

$$x^3 - 13x^2 + 36x = x(x - 9)(x - 4) \; \square$$

Example 2

$$x^4 - 26x^2 + 25 = (x - 5)(x + 5)(x - 1)(x + 1) \; \square$$

In general, a *factor* of a polynomial is any polynomial that divides evenly into the original one. To write a polynomial in *factored form* or to *factor* it is to write it as a product of factors. When the coefficients of the original polynomial are integers we may wish to restrict the factors to polynomials with integral coefficients. In such a case we would expect the coefficients of each factor to have no common integral divisor other than 1 and -1. If we wish to allow nonintegral coefficients, then we can expect no such restriction on the factors. In general, we would like the factors to be as simple as possible. Later in this chapter we will see how the factored form of an expression simplifies computation with algebraic fractions and in later chapters we will see how to solve some equations using a factored form of a polynomial.

The first step in factoring an expression is to look for monomial factors and to rewrite the original polynomial as a product of a monomial and a polynomial in which the terms of the polynomial do not have any common variable factors and, if we restrict the coefficients to integers, the coefficients should have no common integral divisors other than 1 and -1.

Example 3

$$4x^3y^2 + 10x^2y^3 + 6x^4y^4 = 2x^2y^2(2x + 5y + 3x^2y^2)$$

The terms $2x$, $5y$, and $3x^2y^2$ have no common variable factors and 2, 5, and 3 have no common integral divisors except 1 and -1. $\square$

Example 4

$$3ab^3 - 12a^2b^2 + 9a^2b = 3ab(b^2 - 4ab + 3a) \; \square$$

Example 5

$$4p^4q^3 - 12p^2q^5 + 28p^3q^2 = 4p^2q^2(p^2q - 3q^3 + 7p) \; \square$$

Example 6

$$5x^2 + 10x - 15 = 5(x^2 + 2x - 3) \; \square$$

Example 7

$$8y^3 - 20y^2 - 24y = 4y(2y^2 - 5y - 6) \; \square$$

Certain special products appear frequently enough to make knowing them very worthwhile for later work. These were established in Exercise 3 of Section 1.

$$(a + b)^2 = (a + b)(a + b) = a^2 + 2ab + b^2$$
$$(a - b)^2 = (a - b)(a - b) = a^2 - 2ab + b^2$$
$$(a + b)(a - b) = a^2 - b^2$$

Any expression which takes one of these three forms can be easily factored and, therefore, the student should be sure that he remembers them.

Example 8

$$x^2 + 2x + 1 = (x + 1)^2 \ \square$$

Example 9

$$y^2 + 6y + 9 = (y + 3)^2 \ \square$$

Example 10

$$z^2 - 4z + 4 = (z - 2)^2 \ \square$$

Example 11

$$x^2 - 9 = (x + 3)(x - 3) \ \square$$

Example 12

$$y^2 - 1 = (y + 1)(y - 1) \ \square$$

Example 13

$$4x^2 - 9y^2 = (2x + 3y)(2x - 3y) \ \square$$

Example 14

$$9y^2 + 6y + 1 = (3y + 1)^2 \ \square$$

These special forms enable us to "instantly" factor many trinomials containing two perfect squares.

(i) The difference of squares $a^2 - b^2$ can be factored into $(a + b)(a - b)$.
(ii) A polynomial of the form $a^2 + \underline{\hspace{2cm}} + b^2$ is a perfect square if the missing term is either $2ab$ or $-2ab$.

$$a^2 + 2ab + b^2 = (a + b)^2$$
$$a^2 - 2ab + b^2 = (a - b)^2$$

Certainly we cannot expect all trinomials to take one of these forms. What is the general method for factoring trinomials? First, we should realize that many or most trinomials cannot be factored. Our problem is to factor those that can. The first step, remember, is to look for monomial factors.

If we have a polynomial which we are attempting to factor into the form $(a + b)(c + d)$, then we do so by first factoring terms. If our expression is written in decreasing order of powers of some variable, the possibilities become more apparent.

Example 15

Factor

$x^2 - 5x - 6$

There is no common monomial factor other than 1 or -1. If we consider the possible factors of the first and last terms, we have x and x, x^2 and 1, or their negatives for factors of x^2, and 2, -3; -2, 3; 1, -6; and -1, 6 as factors of -6. We can try all the possibilities to see if any work.

$(x^2 + 2)(1 - 3) = (x^2 + 2)(-2) = -2x^2 - 4$

We can see that choosing x^2 and 1 will always yield an expression having no term containing just the first power of x. Thus in order to factor this, we must have an expression of the form $(x \quad)(x \quad)$. Now what are the possibilities? If we use 2 and -3 we obtain

$(x + 2)(x - 3) = x^2 - x - 6$

Using -2 and 3 yields

$(x - 2)(x + 3) = x^2 + x - 6$

Trying -1 and 6 yields

$(x - 1)(x + 6) = x^2 + 5x - 6$

But 1 and -6 yields

$(x + 1)(x - 6) = x^2 - 5x - 6$

which is what we wanted. $\square$

Example 16

Factor the polynomial

$4x^2 + 15x + 9$

There is no common monomial factor (other than 1 or -1 of course). The factors of $4x^2$ are $4x$, x and $2x$, $2x$ and their negatives. The factors of 9 are 9, 1 and 3, 3 and their negatives. Although both $4x^2$ and 9 are perfect squares, the "middle term" is such that this expression is not a perfect square because $(2x \pm 3)^2$ has $\pm 12x$ for a middle term. Thus we rule out $(2x + 3)^2$ and $(2x - 3)^2$ as possibilities. Notice that the middle term has a positive sign. Thus any expression of the form

$$(\quad - \quad) \times (\quad - \quad)$$

is ruled out. We now try the rest of the possibilities:

$$(2x + 1)(2x + 9) = 4x^2 + 20x + 9$$
$$(4x + 1)(x + 9) = 4x^2 + 37x + 9$$
$$(4x + 9)(x + 1) = 4x^2 + 13x + 9$$
$$(4x + 3)(x + 3) = 4x^2 + 15x + 9$$

Thus

$$(4x + 3)(x + 3)$$

is the desired form. $\square$

Factoring expressions is not a step-by-step procedure that you can memorize as you did the multiplication process in arithmetic. You must consider the possibilities and try those that might work. There are, however, some guidelines to follow in factoring:

(i) Look for a "largest" monomial factor and rewrite the original expression as a product of this monomial and a polynomial in order of increasing or decreasing powers of some variable such that the coefficient of the first term is positive.

(ii) Examine the polynomial to see if it fits any special forms such as $a^2 - b^2$, $a^2 + 2ab + b^2$, or $a^2 - 2ab + b^2$.

(iii) If the polynomial is a trinomial, consider all possibilities of the form $(a \quad b)(c \quad d)$ where ac is the first term and bd is the last term. If ac has a positive coefficient, we need only consider a and c with positive coefficients. In considering all possibilities, some can be ruled out by the signs of the terms. If the last term is *positive*, b and d must *agree in sign* and both signs will be the *same as the sign of the middle term*. If the last term has a *negative* sign, b and d will *disagree in sign* and the choice of signs will depend on the middle term.

Example 17

$$3z^3y^2 + 6z^2y^3 + 3zy^4 = 3zy^2(z^2 + 2zy + y^2)$$
$$= 3zy^2(z + y)^2 \;\square$$

Example 18

$$5a^2b^4 - 20a^4b^2 = 5a^2b^2(b^2 - 4a^2)$$
$$= 5a^2b^2(b + 2a)(b - 2a) \;\square$$

In both of these examples, after finding the monomial factor, the remaining polynomial was in one of the special forms.

Example 19

$$4y^2 - 10y - 24 = 2(2y^2 - 5y - 12)$$
$$= 2(2y \quad)(y \quad)$$

The possibilities for second terms pairwise are ± 1, ± 12; ± 2, ± 6; and ± 3, ± 4. We know the signs disagree because the last term has a negative sign. Trying the possibilities we see that

$$2y^2 - 5y - 12 = (2y + 3)(y - 4)$$

Thus

$$4y^2 - 10y - 24 = 2(2y + 3)(y - 4) \quad \square$$

Example 20

$$2x^3 - 9x^2 + 10x = x(2x^2 - 9x + 10)$$
$$= x(2x \quad)(x \quad)$$

The only possibilities for second terms are, pairwise, ± 1, ± 10 and ± 2, ± 5. Since the last term has a positive sign, the second terms agree in sign. Both must have a negative sign because the middle term has a negative sign. Thus

$$2x^3 - 9x^2 + 10x = x(2x - \quad)(x - \quad)$$

and we have yet to try the possibilities. After trying them we see that

$$2x^3 - 9x^2 + 10x = x(2x - 5)(x - 2) \quad \square$$

Certain special products enable us to factor some expressions involving higher powers. The reader should verify the following five products by direct computation. (Some of them were included in Exercise 3, Section 1.)

$$a^3 - b^3 = (a - b)(a^2 + ab + b^2)$$
$$a^3 + b^3 = (a + b)(a^2 - ab + b^2)$$
$$a^4 - b^4 = (a - b)(a^3 + a^2b + ab^2 + b^3) = (a - b)(a + b)(a^2 + b^2)$$
$$a^5 - b^5 = (a - b)(a^4 + a^3b + a^2b^2 + ab^3 + b^4)$$
$$a^5 + b^5 = (a + b)(a^4 - a^3b + a^2b^2 - ab^3 + b^4)$$

In general:

If n is any positive integer, then

$$a^n - b^n = (a - b)(a^{n-1} + a^{n-2}b + a^{n-3}b^2 + \cdots + ab^{n-2} + b^{n-1})$$

If n is an *odd* positive integer, then

$$a^n + b^n = (a + b)(a^{n-1} - a^{n-2}b + a^{n-3}b^2 - \cdots - ab^{n-2} + b^{n-1})$$

These general formulas can be proved formally by using a technique known as mathematical induction but the patterns that appear in the verifications of the first five special products indicate what really happens. We will not offer a proof in this text. In general, these formulas do not provide complete factorization, the second expression can frequently be factored further.

These special products can be used to factor expressions having one of the above forms.

Example 21

$$x^3 - 1 = (x - 1)(x^2 + x + 1) \ \square$$

Example 22

$$x^3 + 8 = (x + 2)(x^2 - 2x + 4) \ \square$$

Example 23

$$y^4 - 16 = (y - 2)(y^3 + 2y^2 + 4y + 8) = (y - 2)(y + 2)(y^2 + 4) \ \square$$

Example 24

$$8x^3 - 27 = (2x - 3)(4x^2 + 6x + 9) \ \square$$

Realize that not all polynomials can be factored into simpler expressions.

Example 25

The polynomial

$$x^2 + 2x + 4$$

cannot be factored.

To see this, suppose

$$x^2 + 2x + 4 = (x + a)(x + b)$$

for some numbers a and b.

Then $a + b = 2$ and $ab = 4$ with a and b positive. In order for $ab = 4$, either $a \geq 2$ or $b \geq 2$. In either case, $a + b$ is larger than 2. Thus no two such numbers exist and $x^2 + 2x + 4$ cannot be factored. $\square$

Example 26

$$y^2 + 9$$

cannot be factored.

To see this, suppose

$$y^2 + 9 = (y + p)(y + q)$$

for some numbers p and q. Then $p + q = 0$ and $pq = 9$. Thus $p = -q$, $pq = -q^2$, and $-q^2 = 9$. Hence $q^2 = -9$ and this is impossible. Thus $y^2 + 9$ cannot be factored. $\square$

EXERCISES

1. Factor.
   (a) $4x^2y^3 - 8x^3y$
   (b) $18x^3y^4 - 9x^2y^5 + 30x^4y^2$
   (c) $20x^2y^3z^4 - 35x^3y^2z^3 + 45x^4y^4z^2$
   (d) $49x^5 + 98x^3 - 343x^7$

2. Factor.
   (a) $x^2 - 4xy + 4y^2$
   (b) $x^2 + 6x + 9$
   (c) $x^2 - 16$
   (d) $y^2 - 2yz + z^2$
   (e) $4z^2 + 4z + 1$
   (f) $4z^2 - 9$
   (g) $25y^2 + 60yx + 36x^2$
   (h) $49x^2 - 42xy + 9y^2$
   (i) $5x^2 - 20$
   (j) $18y^3 - 72y$
   (k) $5y^2 - 30y + 45$
   (l) $24x^2 + 240x + 600$

3. Factor.
   (a) $6a^2 - 5a - 4$
   (b) $6y^2 - 5y - 6$
   (c) $5x^2 + xy - 6y^2$
   (d) $x^2 + 2xy - 8y^2$
   (e) $10x^2 - 9xy - 9y^2$
   (f) $4x^5 - 8x^4 - 4x^3$
   (g) $12x^3y + 2x^2y^2 - 70xy^3$
   (h) $3x^4y^2 - 3x^2y^4$
   (i) $x^2 - 28x + 27$
   (j) $15x^2 + 22x + 8$

4. Factor.
   (a) $20x^2 - 26x - 6$
   (b) $18x^2 - 15xy - 30y^2$
   (c) $21x^2 + xy - 2y^2$
   (d) $20x^2 - 4xy - 3y^2$
   (e) $48x^3 + 92x^2 + 30x$
   (f) $28x^2y + 43xy^2 + 10y^3$
   (g) $10x^2 - 5x - 15$
   (h) $30a^3 - 3a^2b - 3ab^2$
   (i) $45x^2 + 62x + 21$
   (j) $4x^2 + 34xy - 60y^2$

5. Factor.
   (a) $3x^3 - 6x^2 + 105x$
   (b) $12x^2 + 11xy - 5y^2$
   (c) $35a^2 + 2ab - b^2$
   (d) $12x^2 + 142xy + 110y^2$
   (e) $45a^4 - 12a^3 - 12a^2$
   (f) $6x^2 - 7x - 143$
   (g) $80x^2 + 94x + 21$
   (h) $81x^2 - 45x + 6$
   (i) $100x^2 + 100x + 21$
   (j) $63x^2 - 50x - 33$

6. Factor.
   (a) $x^3 + 1$
   (b) $x^4 - 1$
   (c) $x^3 - 27$
   (d) $y^5 - 1$
   (e) $z^5 + 1$
   (f) $x^5 + 32$
   (g) $x^6 - 64$
   (h) $x^7 + 128$

7. Factor.
   (a) $x^3 + 8y^3$
   (b) $a^3 - 125b^3$
   (c) $8y^3 - 27$
   (d) $y^4 - 16$
   (e) $27x^3 + 1$
   (f) $4x^3 + 500$
   (g) $y^5 - 243$
   (h) $2a^3 - 54b^6$
   (i) $8x^3 + 243y^3$
   (j) $1000x^3 + 1$

4. Simplifying Algebraic Fractions

For some reason, not quite known to mathematics educators, many elementary school children are afraid of those terrible numbers known as fractions. A fourth grade boy who is able to add, subtract, multiply, and divide whole numbers with amazing speed and accuracy may have no confidence in his ability to add $\frac{4}{5}$ and $\frac{4}{3}$. A little girl who was "good at math" now decides that "math is not feminine" once she is introduced to fractions. Whatever the cause of this affliction, many students never quite seem to recover. They avoid arithmetic with fractions by memorizing "decimal equivalents" which are, in fact, decimal approximations and then change all fractional problems to decimal problems.

It is the opinion of this author that one basic cause of the difficulty is the false attitude that numbers expressed as fractions are, in some way, different from other numbers. If you examine the axioms for the real number system, you will find that these are laws for *all* numbers, no matter how they are expressed. The techniques for computing with numbers expressed as fractions are given in Theorem 6 of Chapter 1 and depend *only* on the fact that the fraction line is just a division symbol. Thus the principles governing the arithmetic of fractions are just those that govern division of numbers. There is no mystery in the techniques if you remember that a/b just means a divided by b.

The challenge to meet when dealing with algebraic fractions is to express the result in the simplest form. All simplification techniques are based on one fundamental idea.

If a, b, and c are numbers and $b \neq 0$ and $c \neq 0$, then

$$\frac{a}{b} = \frac{ac}{bc}$$

An equivalent phrasing of this principle is:

One can multiply or divide both the numerator and denominator of a fraction by the same nonzero number without changing the value of the fraction.

Example 1

$$\frac{2}{3} = \frac{2 \cdot 5}{3 \cdot 5} = \frac{10}{15} \quad \square$$

Example 2

$$\frac{9}{12} = \frac{3 \cdot 3}{3 \cdot 4} = \frac{3}{4} \quad \square$$

Example 3

$$\frac{.4}{.3} = \frac{4}{3} \;\square$$

Example 4

$$\frac{3x^2}{5x^2} = \frac{3}{5}$$

under the assumption that $x^2 \neq 0$. $\square$

Example 5

$$\frac{(x + 2)(x + 3)}{(x - 7)(x + 3)} = \frac{x + 2}{x - 7}$$

under the assumption that $x + 3 \neq 0$. $\square$

In dealing with fractions in arithmetic, this principle is used in two ways:

(i) To reduce fractions to lowest terms.

Example 6

$$\frac{4}{6} = \frac{2 \cdot 2}{2 \cdot 3} = \frac{2}{3} \;\square$$

(ii) To rewrite summands so that addition or subtraction can be completed using fractions with a common denominator.

Example 7

$$\frac{2}{3} + \frac{4}{5} = \frac{2 \cdot 5}{3 \cdot 5} + \frac{4 \cdot 3}{5 \cdot 3} = \frac{10}{15} + \frac{12}{15} = \frac{22}{15} \;\square$$

We will use this principle in the same ways when dealing with algebraic fractions. In applying the principle to algebraic fractions, we will find factoring a great deal of help because the only expressions that will divide evenly into both the numerator and denominator of a fraction are expressions that are factors of both of them.

Example 8

$$\frac{x^2 - y^2}{xy + y^2} = \frac{(x + y)(x - y)}{(x + y)y} = \frac{x - y}{y} \qquad \text{Dividing numerator and denominator by } x + y. \;\square$$

Example 9

$$\frac{x^2 - 5x + 6}{x^2 - 4} = \frac{(x - 2)(x - 3)}{(x - 2)(x + 2)}$$

Dividing numerator and
denominator by $x - 2$. □

$$= \frac{x - 3}{x + 2}$$

Example 10

$$\frac{a^2 - ab}{ab - b^2} = \frac{a(a - b)}{(a - b)b} = \frac{a}{b}$$

Dividing numerator and
denominator by $a - b$. □

Example 11

$$\frac{(a + b)^4(x - y)^2}{(a + b)^2(x - y)^3} = \frac{(a + b)^2}{x - y}$$

Dividing numerator and
denominator by
$(a + b)^2(x - y)^2$. □

Example 12

$$\frac{4y^2 - 4y - 8}{2y^2 - 12y + 16} = \frac{4(y^2 - y - 2)}{2(y^2 - 6y + 8)}$$

$$= \frac{4(y + 1)(y - 2)}{2(y - 4)(y - 2)}$$

$$= \frac{2(y + 1)}{y - 4}$$

Dividing numerator and
denominator by $2(y - 2)$. □

The general procedure for simplifying algebraic fractions is just the same
as the procedure for reducing numerical fractions to lowest terms:

*Divide the numerator and denominator by all common factors and continue
until no common factors (other than ±1) remain.*

In practice it is usually simplest to factor both the numerator and
denominator as much as possible before trying any division.

After gaining some experience, one does not write down all intermediate
steps but just "keeps track" of the division by "crossing out" or "cancelling"
common factors as he divides by them.

Example 13

$$\frac{x^2 - 5x - 6}{2x^2 - 2} = \frac{(x - 6)(x + 1)}{2(x^2 - 1)} = \frac{(x - 6)\cancel{(x + 1)}}{2(x - 1)\cancel{(x + 1)}} = \frac{x - 6}{2(x - 1)} \quad □$$

Example 14

$$\frac{(a + 1)(a - 2)(a + 3)}{(a - 2)(a - 3)(a + 3)} = \frac{a + 1}{a - 3} \quad \square$$

Example 15

$$\frac{3y^2 - 3}{6y^2 + 12y + 6} = \frac{3(y^2 - 1)}{6(y^2 + 2y + 1)} = \frac{3(y + 1)(y - 1)}{2\cdot 6(y + 1)^{2 1}} = \frac{y - 1}{2(y + 1)} \quad \square$$

In Example 15 we divided the numerator and denominator by 3 and by $(y - 1)$. The method of "keeping track" was to change the exponent on $(y + 1)$ in the denominator rather than cross it out because $(y + 1)$ remains as a factor of the denominator.

Example 16

$$\frac{(a + b)^2(a - b)^{3 2}}{(a + b)^{4 2}(a - b)} = \frac{(a - b)^2}{(a + b)^2} \quad \square$$

Remember that this "crossing out" or "cancelling" is not an algebraic process! It is merely a notational convenience that enables us to eliminate quite a bit of writing. When you are simplifying a fraction the only principle you are using is the fact that you can multiply or divide the numerator and denominator of a fraction by the same nonzero number without changing the value of the fraction.

In Chapter 2 we defined identities to be equations that hold for all possible meaningful substitutions. The equations we write in simplifying algebraic fractions are identities in that if we make meaningful substitutions for variables, all of these equations become true statements.

Example 17

The equation

$$\frac{(x + 3)(x - 2)}{(x + 4)(x - 2)} = \frac{x + 3}{x + 4}$$

yields a true sentence if we substitute any number except 2 or -4 for x. When 2 is substituted for x, the left side becomes meaningless because $0/0$ has no meaning. If $x = -4$, both sides are meaningless. We could be fussy and write

$$\frac{(x + 3)(x - 2)}{(x + 4)(x - 2)} = \frac{x + 3}{x + 4} \qquad \text{for } x \neq 2, -4$$

but we will not do so. Just remember that the equation will hold whenever both sides have meaning and only meaningful substitutions are allowed. ☐

EXERCISES

1. Reduce to lowest terms.
   (a) 14/21   (b) 40/112   (c) 51/85
   (d) 256/1024   (e) 81/135

2. Evaluate the following expressions for $a = 7, b = 5, c = 3$.

   (a) $\dfrac{a^2 + 2ab + b^2}{a^2 - b^2}$

   (b) $\dfrac{a^2 + ac - 6c^2}{a^2 - ac - 2c^2}$

   (c) $\dfrac{b^2 - 1}{b^2 + 3b + 2}$

   (d) $\dfrac{2b^2 + b - 15}{b^2 - b - 12}$

   (e) $\dfrac{a^2 - b^2}{a^3 - b^3}$

   (f) $\dfrac{a^4 - 1}{a^3 - 1}$

3. Simplify.

   (a) $\dfrac{a^2 + 2ab + b^2}{a^2 - b^2}$

   (b) $\dfrac{x^2 - 1}{x^2 + 3x + 2}$

   (c) $\dfrac{x^2 + 5x + 6}{x^2 + 4x + 4}$

   (d) $\dfrac{x^2 + 7x^2 + 12x}{x^2 + 6x + 8}$

   (e) $\dfrac{x^2 + xy - 6y^2}{x^2 - xy - 2y^2}$

   (f) $\dfrac{x^2 - y^2}{x^3 - y^3}$

   (g) $\dfrac{x^2 - 7xy + 10y^2}{x^2 - 5xy + 6y^2}$

   (h) $\dfrac{3x^2 - 12y^2}{15x^2 - 60xy - 60y^2}$

   (i) $\dfrac{x^4 - 1}{x^3 - 1}$

   (j) $\dfrac{x^5 + x^4 + x^3 + x^2 + x + 1}{x^3 + x^2 + x + 1}$

   (k) $\dfrac{2x^2 + x - 15}{x^2 - x - 12}$

   (l) $\dfrac{30x^3 + 81x^2 - 84x}{30x^4 + 6x^3 - 24x^2}$

4. Simplify.

   (a) $\dfrac{x^2 + 2x - 35}{x^2 + 9x + 14}$

   (b) $\dfrac{2x^3 - 7x^2 - 4x}{2x^2 + 7x + 3}$

   (c) $\dfrac{12x^2 + 5x - 2}{4x^3 + 7x^2 - 2x}$

   (d) $\dfrac{3x^3 + 2x^2y - 2xy^2}{3x^2y - 7xy^2 - 2y^3}$

   (e) $\dfrac{6x^2 + 23x + 20}{6x^2 + 19x + 10}$

   (f) $\dfrac{42x^2 - 85x + 42}{42x^2 - 13x - 42}$

   (g) $\dfrac{x^2 - 14xy + 49y^2}{x^2 - 12xy - 14y^2}$

   (h) $\dfrac{x^3 - 1}{x^2 - 6x + 4}$

(i) $\dfrac{a^2 - 9b^2}{a^3 - 27b^3}$

(j) $\dfrac{8y^2 + 26y + 15}{8y^2 - 9}$

(k) $\dfrac{2x^2 + 13xy + 20y^2}{3x^2 + 11xy - 4y^2}$

(l) $\dfrac{a^3 - b^3}{a^2 - b^2}$

5. Multiplication and Division of Algebraic Fractions

Since algebraic expressions are just representations of numbers, the principles governing the arithmetic of algebraic fractions are the same as those governing the arithmetic of numerical fractions. The following basic principles governing multiplication and division were given in Chapter 1.

If a, b, c, and d are numbers such that $b \neq 0$ and $d \neq 0$, then

(i) $\dfrac{a}{b} = \dfrac{ad}{bd}$

(ii) $\dfrac{a}{b} \cdot \dfrac{c}{d} = \dfrac{ac}{bd}$

(iii) $\dfrac{a}{b} \div \dfrac{c}{d} = \dfrac{a}{b} \cdot \dfrac{d}{c} = \dfrac{ad}{bd}$ if $c \neq 0$

We need techniques for dealing with the arithmetic of algebraic fractions that:

1. require a minimum of algebraic computation
2. yield expressions in simplified forms

As you might expect, factoring will play an important role in these techniques. The basic procedure is:

(i) Factor the numerators and denominators of the original fractions.
(ii) Indicate the factored form of the product or quotient.
(iii) Simplify.

Example 1

$$\frac{x + y}{x^2 + 5xy + 6y^2} \cdot \frac{x^2 + 4xy + 3y^2}{x + 4y} = \frac{(x + y)}{(x + 2y)(x + 3y)} \cdot \frac{(x + 3y)(x + y)}{(x + 4y)}$$

$$= \frac{(x + y)(x + 3y)(x + y)}{(x + 2y)(x + 3y)(x + 4y)}$$

$$= \frac{(x + y)^2}{(x + 2y)(x + 4y)} \quad \square$$

Example 2

$$\frac{2x^2 - x - 10}{x^2 - 49} \div \frac{2x^2 + 3x - 2}{x^2 + 14x + 49} = \frac{(2x - 5)(x + 2)}{(x + 7)(x - 7)} \div \frac{(2x - 1)(x + 2)}{(x + 7)^2}$$

$$= \frac{(2x - 5)(x + 2)}{(x + 7)(x - 7)} \cdot \frac{(x + 7)^2}{(2x - 1)(x + 2)}$$

$$= \frac{(2x - 5)\cancel{(x + 2)}(x + 7)^{\cancel{2}1}}{\cancel{(x + 7)}(x - 7)(2x - 1)\cancel{(x + 2)}}$$

$$= \frac{(2x - 5)(x + 7)}{(x - 7)(2x - 1)}$$

$$= \frac{2x^2 + 9x - 35}{2x^2 - 15x + 7} \quad \square$$

Multiplication of three or more fractions is accomplished in the same fashion.

Example 3

$$\frac{x^2 - 1}{x^2 - 4} \cdot \frac{x^2 + 6x + 8}{x^2 + 9x + 8} \cdot \frac{x^3 - 8}{x^3 - 1}$$

$$= \frac{(x + 1)(x - 1)}{(x + 2)(x - 2)} \cdot \frac{(x + 4)(x + 2)}{(x + 1)(x + 8)} \cdot \frac{(x - 2)(x^2 + 2x + 4)}{(x - 1)(x^2 + x + 1)}$$

$$= \frac{\cancel{(x + 1)}\cancel{(x - 1)}(x + 4)\cancel{(x + 2)}\cancel{(x - 2)}(x^2 + 2x + 4)}{\cancel{(x + 2)}\cancel{(x - 2)}\cancel{(x + 1)}(x + 8)\cancel{(x - 1)}(x^2 + x + 1)}$$

$$= \frac{(x + 4)(x^2 + 2x + 4)}{(x + 8)(x^2 + x + 1)}$$

This cannot be simplified further. If we complete the indicated multiplication, we obtain

$$\frac{x^3 + 6x^2 + 12x + 16}{x^3 + 9x^2 + 9x + 8} \quad \square$$

Frequently the indicated product of factored expressions can be simplified without rewriting as one fraction.

Example 4

$$\frac{2x^2 - 5x - 12}{12x^2 + 8x + 1} \cdot \frac{6x^2 + 19x + 3}{2x^2 + x - 3} = \frac{\cancel{(2x + 3)}(x - 4)}{\cancel{(6x + 1)}(2x + 1)} \cdot \frac{\cancel{(6x + 1)}(x + 3)}{\cancel{(2x + 3)}(x - 1)}$$

$$= \frac{(x - 4)(x + 3)}{(2x + 1)(x - 1)} = \frac{x^2 - x - 12}{2x^2 - x - 1} \quad \square$$

Example 5

$$\frac{6x^2 - 5x + 1}{x^2 + 4x + 4} \div \frac{3x^2 - 16x + 5}{x^2 - 4}$$

$$= \frac{(3x - 1)(2x - 1)}{(x + 2)^{2}{}_1} \cdot \frac{(x + 2)(x - 2)}{(3x - 1)(x - 5)} \qquad \text{(factoring and inverting the divisor in the same step)}$$

$$= \frac{2x^2 - 5x + 2}{x^2 - 3x - 10} \quad \square$$

Fractions in which the numerator and/or denominator are not just polynomials but rather more complex expressions are called *complex fractions*. Since the fraction line just means division, we can write such expressions as indicated divisions and complete the computation.

Example 6

$$\frac{\dfrac{x^2 + 3xy + 2y^2}{x^2 + 5xy - 6y^2}}{\dfrac{x^2 + 4xy + 4y^2}{x^2 - 36y^2}}$$

is a complex fraction.

We can rewrite it using division notation. Thus,

$$\frac{\dfrac{x^2 + 3xy + 2y^2}{x^2 + 5xy - 6y^2}}{\dfrac{x^2 + 4xy + 4y^2}{x^2 - 36y^2}} = \frac{x^2 + 3xy + 2y^2}{x^2 + 5xy - 6y^2} \div \frac{x^2 + 4xy + 4y^2}{x^2 - 36y^2}$$

$$= \frac{(x + y)(x + 2y)}{(x + 6y)(x - y)} \cdot \frac{(x + 6y)(x - 6y)}{(x + 2y)^{2}{}_1}$$

$$= \frac{(x + y)(x - 6y)}{(x - y)(x + 2y)} = \frac{x^2 - 5xy - 6y^2}{x^2 - xy - 2y^2} \quad \square$$

Example 7

$$\frac{\dfrac{a^2 + 10ab + 16b^2}{a^2 - 7ab + 10b^2}}{\dfrac{a^2 - 4b^2}{a^2 - 2ab - 15b^2}} = \frac{a^2 + 10ab + 16b^2}{a^2 - 7ab + 10b^2} \div \frac{a^2 - 4b^2}{a^2 - 2ab - 15b^2}$$

$$= \frac{(a + 2b)(a + 8b)}{(a - 5b)(a - 2b)} \cdot \frac{(a - 5b)(a + 3b)}{(a + 2b)(a - 2b)}$$

$$= \frac{(a + 8b)(a + 3b)}{(a - 2b)^2} = \frac{a^2 + 11ab + 24b^2}{a^2 - 4ab + 4b^2} \quad \square$$

EXERCISES

1. Complete the indicated computation and simplify.

(a) $\dfrac{x^2}{x+2} \cdot \dfrac{x+2}{x^3}$

(b) $\dfrac{x^2-4}{x+3} \cdot \dfrac{x^2+5x+6}{x-2}$

(c) $\dfrac{x^2+xy-6y^2}{6x^2-xy-y^2} \cdot \dfrac{2x-y}{x-2y}$

(d) $\dfrac{x^2+3x+9}{3x+9} \cdot \dfrac{x^2-9}{x^3-27}$

(e) $\dfrac{x}{x+1} \div \dfrac{x}{x-1}$

(f) $\dfrac{2x}{x^2-1} \div \dfrac{x^2}{x-1}$

(g) $\dfrac{x+1}{x} \div \dfrac{x-1}{2x}$

(h) $(x^2-4y^2) \div \dfrac{x+2y}{x-2y}$

(i) $\dfrac{2x}{x^2-4x-5} \div \dfrac{3x}{x^2-1}$

(j) $\dfrac{5}{x^2-1} \div \dfrac{3}{x^3-1}$

(k) $\dfrac{a^2-7ab+12b^2}{a^3-8a^2b-200b^2} \cdot \dfrac{a^2-7ab+10b^2}{a^3-2a^2b-8ab^2}$

(l) $\dfrac{x^3}{x^3-1} \div \dfrac{x^2}{x^2-1}$

(m) $\dfrac{4x^2+12x+9}{x^2+11x-12} \div \dfrac{x^3+27}{x^3+1}$

(n) $\dfrac{x^2+5x-6}{x^2+9x+8} \div \dfrac{x^2+7x+6}{x^2+10x+16}$

2. Complete the following computations and simplify.

(a) $\dfrac{x^2-4}{x^2+7x+12} \cdot \dfrac{x^2-9}{x^2+4x+4} \cdot \dfrac{x^2+5x+4}{x^2-6x+9}$

(b) $\dfrac{x^3-2x^2y+xy^2}{x^2y-y^3} \cdot \dfrac{x^3-y^3}{x^2-xy} \cdot \dfrac{xy+y^2}{x^2+xy+y^2}$

(c) $\dfrac{a^2-b^2}{a^2+b^2} \cdot \dfrac{a+b}{(a-b)^2} \cdot \dfrac{a^2-b^2}{(a+b)^2}$

(d) $\dfrac{x^3-y^3}{x^3} \cdot \dfrac{x^2-y^2}{(x-y)^2} \cdot \dfrac{x^2+2xy+y^2}{x^2y+xy^2+y^3}$

(e) $\dfrac{x^2+6x+8}{x^2-9} \cdot \dfrac{x^2+7x+10}{x^2-16} \cdot \dfrac{x^2+7x+12}{x^2+4x+4}$

3. Simplify.

(a) $\dfrac{\dfrac{x+3}{x^2}}{\dfrac{x-7}{x}}$

(b) $\dfrac{\dfrac{x^2-4}{x+3}}{\dfrac{x+2}{x^2-9}}$

(c) $\dfrac{\dfrac{a^2+6ab+9b^2}{a^2+4ab+4b^2}}{\dfrac{a^2-9b^2}{a^2-4b^2}}$

(d) $\dfrac{\dfrac{x^2-1}{x^2+5x+6}}{\dfrac{x^2-4}{x^2+4x+3}}$

(e) $\dfrac{\dfrac{x^2-16}{x^2+7x-8}}{\dfrac{x^2-7x+12}{x^2-6x-16}}$

(f) $\dfrac{\dfrac{x^4-16}{x^2-4x+5}}{\dfrac{x^2+4}{x^2-25}}$

6. Addition and Subtraction of Algebraic Fractions

Since algebraic expressions are just representations of numbers, the principles governing addition and subtraction of algebraic fractions are the same as those governing addition and subtraction of numerical fractions and were given in Chapter 1.

If a, b, c, and d are numbers such that $b \neq 0$ and $d \neq 0$, then

(i) $\dfrac{a}{b} = \dfrac{ad}{bd}$

(ii) $\dfrac{a}{b} + \dfrac{c}{b} = \dfrac{(a + c)}{b}$

(iii) $\dfrac{a}{b} - \dfrac{c}{b} = \dfrac{(a - c)}{b}$

(iv) $\dfrac{a}{b} + \dfrac{c}{d} = \dfrac{(ad + bc)}{bd}$

(v) $\dfrac{a}{b} - \dfrac{c}{d} = \dfrac{(ad - bc)}{bd}$

Again *we need techniques that require a minimum of algebraic computation and yield expressions in simplified forms.* As before, factoring will play an important role.

If we add two numerical fractions, we usually find a "least common denominator," apply principle (i) above to rewrite the given fractions with this common denominator, and then use principle (ii) to complete our computation.

Example 1

$$\frac{5}{8} + \frac{3}{20} = \frac{5 \cdot 5}{8 \cdot 5} + \frac{3 \cdot 2}{20 \cdot 2} = \frac{25}{40} + \frac{6}{40} = \frac{31}{40} \quad \square$$

We can apply the same principles to add algebraic fractions although a "least common denominator" has no meaning. In this case we want a "least *complicated* common denominator."

Example 2

$$\frac{4}{xy^2} + \frac{3}{x^2y} = \frac{4x}{x \cdot xy^2} + \frac{3y}{x^2y \cdot y}$$

$$= \frac{4x}{x^2y^2} + \frac{3y}{x^2y^2}$$

$$= \frac{4x + 3y}{x^2y^2}$$

Multiplying the numerator and denominator of the first fraction by x and the numerator and denominator of the second by y yields two fractions with x^2y^2 as the denominator. $\square$

In arithmetic, our numerical knowledge enables us, in most cases, to find a least common denominator. If finding such a denominator is too complicated, we can just apply principle (iv) which, in fact, uses the product of the denominators as the common denominator although, in general, it is not the smallest common denominator.

Example 3

Add

$$\frac{2}{843} + \frac{5}{438}$$

In this case, a great deal of time would be spent finding a least common denominator whereas principle (iv) enables us to proceed directly with less time consumed.

$$\frac{2}{843} + \frac{5}{438} = \frac{2(438) + (843)5}{843 \cdot 438}$$

$$= \frac{876 + 4215}{369,234}$$

$$= \frac{5091}{369,234} \quad \square$$

At this point we might wish to attempt to reduce the fraction to lowest terms but such an effort is not always necessary or desirable.

In adding algebraic fractions, we can use either principle (ii) or principle (iv). If a convenient common denominator is available, we will use (ii) and avoid a goodly amount of computation. Similarly, subtraction can be performed using (iii) if a convenient common denominator is available or using (v) if not. We usually prefer fractions simplified as much as possible and using the "least complicated denominator" saves work in simplification.

Example 4

$$\frac{x + 2y}{x^2 - y^2} + \frac{x - 3y}{x^2 + 2xy + y^2} = \frac{x + 2y}{(x + y)(x - y)} + \frac{x - 3y}{(x + y)^2}$$

$$= \frac{(x + 2y)(x + y)}{(x + y)^2(x - y)} + \frac{(x - 3y)(x - y)}{(x + y)^2(x - y)}$$

$$= \frac{(x^2 + 3xy + 2y^2) + (x^2 - 4xy + 3y^2)}{(x + y)^2(x - y)}$$

$$= \frac{2x^2 - xy + 5y^2}{(x + y)^2(x - y)} \quad \square$$

In this example we found a common denominator by first factoring the original denominators and then choosing as our common denominator the polynomial (in factored form) which contained as factors all the polynomials that appear as factors in the denominator. This is, in fact, the general technique used in adding or subtracting algebraic fractions.

(a) Factor all denominators.

(b) Choose as a common denominator the product of all factors appearing in any of the denominators using each factor to the highest power that it appears.

This product is certainly divisible by each denominator and thus is a common denominator. Furthermore, no less complicated denominator would do since each factor of the original denominator must divide the common denominator and, therefore, must be a factor of any common denominator. Now we need only apply principle (i) and then (ii) or (iii) to complete our computation.

Example 5

$$\frac{4x^2 + 2x - 1}{x^2 + 2x - 15} - \frac{3x - 7}{x^2 + 4x - 21}$$

$$= \frac{4x^2 + 2x - 1}{(x - 3)(x + 5)} - \frac{3x - 7}{(x - 3)(x + 7)}$$

$$= \frac{(4x^2 + 2x - 1)(x + 7)}{(x - 3)(x + 5)(x + 7)} - \frac{(3x - 7)(x + 5)}{(x - 3)(x + 5)(x + 7)}$$

$$= \frac{4x^3 + 30x^2 + 13x - 7}{(x - 3)(x + 5)(x + 7)} - \frac{3x^2 + 8x - 35}{(x - 3)(x + 5)(x + 7)}$$

$$= \frac{(4x^3 + 30x^2 + 13x - 7) - (3x^2 + 8x - 35)}{(x - 3)(x + 5)(x + 7)}$$

$$= \frac{4x^3 + 27x^2 + 5x + 28}{(x - 3)(x + 5)(x + 7)} \quad \square$$

At this point we can leave the expression in this form or else complete the indicated multiplication in the denominator. We also could try to simplify if we feel that it is needed. Realize that we only need to check to see if any of the factors in the denominator will divide the numerator. In order for $(x - 3)$ or $(x + 5)$ to divide the numerator, 3 or 5 would have to divide 28. Thus the only possibility left is $(x + 7)$ and division shows that this is not a factor.

Addition of three algebraic fractions is accomplished by the same method; we need only to factor all three denominators and then choose the common denominator as before.

Example 6

$$\frac{4x + 2}{x^2 - x - 2} + \frac{x^2 - 5}{x^2 - 6x - 7} + \frac{2x + 3}{x^2 - 4}$$

$$= \frac{4x + 2}{(x - 2)(x + 1)} + \frac{x^2 - 5}{(x + 1)(x - 7)} + \frac{2x + 3}{(x + 2)(x - 2)}$$

$$= \frac{(4x + 2)(x - 7)(x + 2)}{(x - 2)(x + 1)(x - 7)(x + 2)} + \frac{(x^2 - 5)(x + 2)(x - 2)}{(x - 2)(x + 1)(x - 7)(x + 2)}$$

$$+ \frac{(2x + 3)(x + 1)(x - 7)}{(x - 2)(x + 1)(x - 7)(x + 2)}$$

$$= \frac{(4x^3 - 18x^2 - 66x - 28) + (x^4 - 9x^2 + 20) + (2x^3 - 9x^2 - 32x - 21)}{(x - 2)(x + 1)(x - 7)(x + 2)}$$

$$= \frac{x^4 + 6x^3 - 36x^2 - 98x - 29}{(x - 2)(x + 1)(x - 7)(x + 2)} \quad \square$$

In order to see if this can be simplified, we need only to check to see if any factors in the denominator are factors of the numerator. Division shows that they are not and the result cannot be simplified.

EXERCISES

1. Complete the indicated computation and reduce to lowest terms.

(a) $\frac{2}{3} + \frac{5}{6}$ (b) $\frac{3}{14} + \frac{5}{6}$

(c) $\frac{7}{8} - \frac{5}{12}$ (d) $\frac{7}{20} + \frac{5}{16}$

(e) $\frac{2}{7} - \frac{5}{3}$ (f) $\frac{5}{27} + \frac{8}{12}$

(g) $\frac{11}{54} + \frac{15}{40}$ (h) $\frac{19}{94} + \frac{13}{35}$

2. Complete the indicated computations and simplify.

(a) $\dfrac{3x^2}{4y^2} + \dfrac{3y}{2x}$ (b) $\dfrac{5ab^3}{6c^2} - \dfrac{3c^3}{4b^2}$

(c) $\dfrac{2}{x} + \dfrac{3}{xy} + \dfrac{7}{y}$ (d) $\dfrac{4}{b^2} + \dfrac{5}{d^2} + \dfrac{9}{bd}$

(e) $\dfrac{x + y}{x^2} + \dfrac{2x - y}{y^2} + \dfrac{3x + y}{xy}$ (f) $\dfrac{2x - 1}{x^2} + \dfrac{5x + 2}{x^3} - \dfrac{3x + 1}{x}$

(g) $\dfrac{x - 7y}{x^2y} - \dfrac{3x + 2y}{xy^2}$ (h) $\dfrac{x + 3y}{x^2y^2} + \dfrac{5x^2 - 2y}{x^3y} + \dfrac{3x + y}{xy^2}$

3. Complete the indicated computation and simplify.

(a) $\dfrac{x}{x+1} + \dfrac{x}{x-1}$

(b) $\dfrac{2x}{x^2-1} - \dfrac{1}{x-1}$

(c) $\dfrac{x^2+xy-6y^2}{6x^2-xy-y^2} + \dfrac{3x+5y}{2x-y}$

(d) $\dfrac{x^2+5x}{x^2+6x+9} - \dfrac{x^2+2}{x^3+27}$

(e) $\dfrac{x+1}{x} + \dfrac{x-1}{2x}$

(f) $(x^2+4y^2) \cdot \dfrac{x+2y}{x+y} - \dfrac{x+2y}{x-2y}$

(g) $\dfrac{2x}{x^2-4x-5} + \dfrac{3x}{x^2-1}$

(h) $\dfrac{5}{x^2-1} - \dfrac{3}{x^3-1}$

(i) $\dfrac{x^2+5x-7}{x^2-x-6} + \dfrac{x^2+3x-2}{x^2+5x+6}$

(j) $\dfrac{x^2+3xy+5y^2}{x^2+3xy-10y^2} + \dfrac{x^2-7xy+y^2}{x^2-4y^2}$

(k) $\dfrac{a^3+2a^2b+b^2}{a^3-b^3} - \dfrac{a+3b}{a^2-b^2}$

(l) $\dfrac{a^2+4ab-5b^2}{a^2+4ab+3b^2} - \dfrac{a^2-3ab}{a^2+7ab+12b^2}$

4. Complete the indicated computations and simplify.

(a) $\dfrac{2x+3}{x^2+7x+10} + \dfrac{4x-1}{x^2+5x+6} + \dfrac{5x-7}{x^2+8x+15}$

(b) $\dfrac{4}{x-1} + \dfrac{5}{x} + \dfrac{6}{x+1}$

(c) $\dfrac{3}{x^2+2x+1} + \dfrac{4}{x^2-2x+1} + \dfrac{5}{x^2-1}$

(d) $\dfrac{2x-3}{x^2-3x-4} + \dfrac{x^2+1}{x^2-16} + \dfrac{x^2-3}{x^2+8x+16}$

(e) $\dfrac{2x-1}{x+2} + \dfrac{x-1}{x-2} + \dfrac{x^2-3x}{x^2-4} + \dfrac{x^2-2x-3}{(x-2)^2}$

(f) $\dfrac{2x+1}{x^2-4} - \dfrac{x^2-1}{x-2} - \dfrac{x-3}{x+2}$

(g) $\dfrac{x^2+3x-4}{x^2-3x-4} + \dfrac{x^2+5x+6}{x^2+4x+3} - \dfrac{x^2+5}{x^2+8x+16}$

(h) $\dfrac{x}{x^2-y^2} + \dfrac{y}{x^3-y^3} + \dfrac{xy}{x^3+y^3}$

5. Complete the following computations and simplify. (First work inside the parentheses.)

(a) $\left(x-\dfrac{1}{x}\right) \div \left(x+\dfrac{1}{x}\right)$

(b) $\left(\dfrac{x}{y}+1\right) \div \left(\dfrac{x^2}{y^2}-1\right)$

(c) $\left(\dfrac{x}{2} - \dfrac{2}{x}\right) \cdot \left(4 - \dfrac{2x}{x+2}\right)$

(d) $\left(\dfrac{a+b}{a-b} - \dfrac{a^3+b^3}{a^3-b^3}\right) \div \left(\dfrac{a+b}{a-b} - \dfrac{a^2+b^2}{a^2-b^2}\right)$

*7. The Factor Theorem

In this section we develop a rather strong theorem which provides a test for determining whether or not a particular binomial of the form $x - a$ is a factor of a given polynomial. This gives us one of the few methods of factoring polynomials such as

$$x^4 + x^3 - 4x^2 + x - 10$$

The polynomial $a_n x^n + a_{n-1} x^{n-1} + \cdots + a_1 x + a_0$ where each a_i is a real number is said to have *degree* n providing $a_n \neq 0$. Intuitively, the degree of a polynomial in one variable, say x, is the highest power of x appearing with a nonzero coefficient. We will frequently find it convenient to denote a polynomial in one variable x by symbols such as $f(x)$, $g(x)$, and $p(x)$. If c is a number and $f(x)$ is a polynomial, then $f(c)$ is the number obtained by substituting c for x in the polynomial. The expression $f(x)$ is read "f of x" and $f(c)$ is read "f of c." To compute $f(c)$ is to "evaluate $f(x)$ at c."

Example 1

If

$$f(x) = x^2 - 3x + 5$$

then

$$f(2) = 4 - 6 + 5 = 3$$

and

$$f(7) = 49 - 21 + 5 = 33 \quad \square$$

Example 2

If

$$g(x) = 3x^3 - 2x^2 + 3x - 7$$

then

$$g(2) = 3(8) - 2(4) + 3(2) - 7 = 15$$

and

$$g(1) = -3 \quad \square$$

If we have two polynomials and try to divide the first by the second, either it divides evenly or the long division technique will produce a remainder of degree less than that of the second.

Example 3

$$
\begin{array}{r}
4x^3 - 11x^2 + 28x - 60 \\
x^2 + 2x - 1 \overline{\smash{)}4x^5 - 3x^4 + 2x^3 + 7x^2 + 5x - 4} \\
\underline{4x^5 + 8x^4 - 4x^3} \\
-11x^4 + 6x^3 + 7x^2 \\
\underline{-11x^4 - 22x^3 + 11x^2} \\
28x^3 - 4x^2 + 5x \\
\underline{28x^3 + 56x^2 - 28x} \\
-60x^2 + 33x - 4 \\
\underline{-60x^2 - 120x + 60} \\
153x - 64
\end{array}
$$

The remainder is $153x - 64$ which is of degree less than the degree of $x^2 + 2x - 1$. $\square$

We state this principle more formally as follows:

Division Algorithm for Polynomials

If $f(x)$ and $g(x)$ are polynomials in the one variable x with $g(x) \neq 0$, then there exist unique polynomials $q(x)$ and $r(x)$ such that

$$f(x) = g(x)q(x) + r(x)$$

and $r(x) = 0$ or the degree of $r(x)$ is less than the degree of $g(x)$.

Notice that if $g(x)$ is of degree 1, then $r(x)$ is either 0 or a polynomial of degree 0, that is, just a number. Thus if a is a number and $f(x)$ a polynomial, there exists a polynomial $q(x)$ and a number r such that

$$f(x) = (x - a)q(x) + r$$

Then

$$
\begin{aligned}
f(a) &= 0 \cdot q(a) + r \\
&= 0 + r \\
&= r
\end{aligned}
$$

Thus the remainder after dividing $f(x)$ by $(x - a)$ is just $f(a)$. Furthermore, since $(x - a)$ is a factor of $f(x)$ if and only if $r = 0$, $(x - a)$ is a factor of $f(x)$ if and only if $f(a) = 0$.

These two results are usually stated more formally as follows:

THE REMAINDER THEOREM

If $f(x)$ is a polynomial and $a \in R$, then the remainder after dividing $f(x)$ by $(x - a)$ is $f(a)$.

THE FACTOR THEOREM

If $f(x)$ is a polynomial and $a \in R$, $(x - a)$ is a factor of $f(x)$ if and only if $f(a) = 0$.

Example 4

$x - 2$ is not a factor of $g(x) = 2x^2 + 3x - 4$ because $g(2) = 10 \neq 0$. $\square$

Example 5

$x - 2$ is a factor of $f(x) = x^3 + 5x^2 - 3x - 22$ because $f(2) = 0$. If we divide $x^3 + 5x^2 - 3x - 22$ by $x - 2$ we obtain a quotient of $x^2 + 7x + 11$ and, therefore,

$$x^3 + 5x^2 - 3x - 22 = (x - 2)(x^2 + 7x + 11)$$

Further attempts to factor $x^2 + 7x + 11$ with integral coefficients will fail because there are no two positive integers whose product is 11 and whose sum is 7. $\square$

These theorems can be used to determine the factors of a polynomial which are of the form $(x - a)$. If $f(a) = 0$, $(x - a)$ is a factor and we can factor simply by dividing. If $f(x)$ is a polynomial with integral factors and we are looking only for factors with integral coefficients, then in order for $(x - a)$ to be a factor, a must divide the constant term of $f(x)$, that is, if $f(x) = b_n x^n + b_{n-1} x^{n-1} + \cdots + b_1 x + b_0$ and $(x - a)$ is a factor of $f(x)$, then a must be a factor of b_0.

EXERCISES

1. Let $p(x) = 4x^5 - 3x^3 + 5x^2 - 7x + 2$.
   Compute $p(0)$, $p(1)$, $p(-2)$, and $p(3)$.
2. Let $f(x) = x^2 - 3x - 4$.
   (a) Compute $f(1), f(-1), f(2), f(-2), f(4), f(-4)$.
   (b) Can you factor $f(x)$?
3. Let $g(x) = x^3 + 5x^2 - 13x - 28$.
   (a) Compute $g(1), g(-1), g(2), g(-2), g(4), g(-4), g(7), g(-7), g(14), g(-14)$.
   (b) Can you factor $g(x)$?
4. Let $h(x) = x^4 + 9x^3 + 21x^2 - x - 30$.
   (a) Compute $h(1), h(-1), h(2), h(-2), h(3), h(-3), h(5), h(-5), h(6), h(-6)$.
   (b) Can you factor $h(x)$?

5. Let $q(x) = x^4 + 4x^3 - 7x^2 - 24x + 24$.
   (a) Compute $q(1)$, $q(-1)$, $q(2)$, $q(-2)$, $q(3)$, $q(-3)$, $q(4)$, $q(-4)$.
   (b) Can you factor $q(x)$?

Factor the following polynomials using the factor theorem. Factor as much as possible.

6. $x^3 - 12x^2 + 41x - 42$ 7. $x^3 + 6x^2 + 21x + 36$
8. $x^3 - 3x^2 - 3x - 4$ 9. $x^5 - 9x^3 - 8x^2 + 72$
10. $x^5 + 5x^4 - 6x^3 - 27x^2 - 135x + 162$

4

Exponents, Roots, and Radicals

In the first three chapters we have been concerned with algebraic expressions that can be formed using numerals, variables, and the four arithmetic operations of addition, subtraction, multiplication, and division. In this chapter we will be concerned with certain types of irrational numbers and with algebraic expressions related to such numbers. These numbers arise quite naturally in practical problems and the ability to deal with them is important to all students of algebra and its applications. As we discuss these numbers, we will find it convenient to extend the meaning of exponents in order to simplify many expressions and computations.

1. Integral Exponents

Remember that our original purpose for defining exponents was to provide an abbreviation for products in which all factors are the same. Thus far only positive integers have been used as exponents. We can give meaning to all integral exponents but natural meanings such as $a^2 = a \cdot a$ are not obviously available to us. We will give meaning to zero exponents and negative integral exponents in such a way as to extend the first law of exponents and, in doing so, we will see that all four laws are extended. For convenience, the four laws are listed below.

117

If a and b are real numbers and m and n are positive integers, then

(i) $a^n a^m = a^{n+m}$
(ii) $(a^m)^n = a^{mn}$
(iii) $(ab)^n = a^n b^n$
(iv) If $b \neq 0$,

$$\left(\frac{a}{b}\right)^n = \frac{a^n}{b^n}$$

What meaning should we assign to expressions such as 2^0, 5^0, $(-4)^0$, and $(.1)^0$? If the first law of exponents is to be extended, we would expect to have

$$2^3 \cdot 2^0 = 2^{3+0} = 2^3$$

that is,

$$2^3 \cdot 2^0 = 2^3$$

Also we would expect

$$5 \cdot 5^0 = 5^1 \cdot 5^0 = 5^{1+0} = 5^1$$

that is,

$$5 \cdot 5^0 = 5$$
$$(-4)^0 \cdot (-4)^7 = (-4)^{0+7} = (-4)^7$$

that is,

$$(-4)^0 \cdot (-4)^7 = (-4)^7$$

and

$$(.1)^0 (.1)^{43} = (.1)^{0+43} = (.1)^{43}$$

that is,

$$(.1)^0 (.1)^{43} = (.1)^{43}$$

In the cases above, in order for the equations to be true we must have

$$2^0 = 1, \quad 5^0 = 1, \quad (-4)^0 = 1 \quad \text{and} \quad (.1)^0 = 1$$

This suggests the following general definition:

If a is a real number, then $a^0 = 1$.

Thus $9^0 = 1$, $(-17)^0 = 1$, $\pi^0 = 1$, and $0^0 = 1$. (Some authors prefer to leave 0^0 undefined but since trouble arises only when considering limits, no difficulties will arise in this course. It is convenient to consider $x^0 = 1$ for any number x.)

Now that we have a meaning for zero exponents, how can we give meaning to negative integral exponents? Again we wish to extend the first law of exponents, that is, we would expect $a^x a^y = a^{x+y}$ to hold for any integers x and y. In order for this law to extend we must have

$$5^2 \cdot 5^{-2} = 5^{2+(-2)} = 5^0 = 1$$
$$3^5 \cdot 3^{-5} = 3^{5+(-5)} = 3^0 = 1$$

and

$$(-.04)^3 (-.04)^{-3} = (-.04)^{3+(-3)} = (-.04)^0 = 1$$

Thus in order to extend the first law of exponents, we must have

$$5^{-2} = \frac{1}{5^2}, \quad 3^{-5} = \frac{1}{3^5} \quad \text{and} \quad (-.04)^{-3} = \frac{1}{(-.04)^3}$$

This suggests the following general definition:

If a is a *nonzero* real number and n is a positive integer, then

$$a^{-n} = \frac{1}{a^n}$$

Thus

$$a^n \cdot a^{-n} = a^{n+(-n)} = a^0 = 1$$

because

$$a^n \cdot \left(\frac{1}{a^n}\right) = 1$$

Notice that if $a \neq 0$ and y is any integer, then

$$a^y a^{-y} = 1$$

because

$$a^{-y} = \frac{1}{a^y} \qquad \text{if } y > 0$$

$$a^y = \frac{1}{a^{-y}} \qquad \text{if } y < 0$$

and

$$a^y = a^{-y} = 1 \qquad \text{if } y = 0$$

Thus

$$a^{-y} = 1/a^y$$

for any integer y if $a \neq 0$.

Note that 0^{-n} has no meaning if n is a positive integer.

Our definitions were made to extend the first law of exponents,

$a^n a^m = a^{n+m}$, at least for the case where $n = 0$ or $n = -m$. If we were successful, then *all four laws* can be extended for integral exponents. We see now that this is, in fact, the case.

In order to extend (i) we need to show:

I. If b is a nonzero number with x and y integers, then $b^x \cdot b^y = b^{x+y}$.

Proof of I:

Case 1

If $x > 0$ and $y > 0$, the equation holds by (i).

Case 2

If $x < 0$ and $y < 0$, then $x + y < 0$ and

$$b^x \cdot b^y = \frac{1}{b^{-x}} \cdot \frac{1}{b^{-y}}$$

$$= \frac{1}{b^{-x} \cdot b^{-y}}$$

$$= \frac{1}{b^{(-x)+(-y)}} \qquad \text{by (i) because } -x > 0 \text{ and } -y > 0$$

$$= \frac{1}{b^{-(x+y)}}$$

$$= b^{x+y}$$

Case 3

If $x < 0$, $y > 0$, and $x + y > 0$, then $-x > 0$ and by (i) we have

$$b^{-x} \cdot b^{(x+y)} = b^{(-x)+(x+y)} = b^y$$

that is,

$$b^{-x} \cdot b^{x+y} = b^y$$

Thus

$$b^x b^{-x} \cdot b^{x+y} = b^x b^y \qquad \text{(Multiplying both sides by } b^x.)$$

$$1 \cdot b^{x+y} = b^x b^y \qquad \left(\text{because } b^x b^{-x} = b^x \cdot \frac{1}{b^x} = 1 \right)$$

and

$$b^x b^y = b^{x+y}$$

Case 4

If $x < 0$, $y > 0$, and $x + y < 0$, then $-y < 0$ and, by Case 2,

$$b^{x+y} \cdot b^{-y} = b^{(x+y)+(-y)} = b^x$$

that is,

$$b^{x+y}b^{-y} = b^x$$
$$b^{x+y}b^{-y}b^y = b^x b^y \qquad \text{(Multiplying both sides by } b^y.)$$
$$b^{x+y} \cdot 1 = b^x b^y \qquad \left(\text{because } b^{-y}b^y = \frac{1}{b^y} \cdot b^y = 1\right)$$

and

$$b^x b^y = b^{x+y}$$

The case where $x > 0$, $y < 0$, and $x + y > 0$ and the case where $x > 0$, $y < 0$, and $x + y < 0$ are essentially the same as Cases 3 and 4 because multiplication is commutative. Thus we see that I holds and (i) is extended.

Example 1

$$5^3 \cdot 5^{-4} = 5^{-1} \ \square$$

Example 2

$$(17)^{15}(17)^{-6} = (17)^9 \ \square$$

Example 3

$$(.3)^{-2}(.3)^{-4} = (.3)^{-6} \ \square$$

In order to extend the second law of exponents, we need to show:

II. If b is a nonzero real number with x and y integers, then

$$(b^x)^y = b^{xy}$$

Proof of II:

If $x > 0$ and $y > 0$, $(b^x)^y = b^{xy}$ by (i).
If $x = 0$, $(b^0)^y = 1^y = 1 = b^0 = b^{0 \cdot y}$.
If $y = 0$, $(b^x)^0 = 1 = b^0 = b^{x \cdot 0}$.
If $x > 0$ and $y < 0$,

$$(b^x)^y = \frac{1}{(b^x)^{-y}} = \frac{1}{b^{x(-y)}} = \frac{1}{b^{-(xy)}} = b^{xy}$$

If $x < 0$ and $y > 0$,

$$(b^x)^y = \left(\frac{1}{b^{-x}}\right)^y = \frac{1^y}{(b^{-x})^y} = \frac{1}{b^{(-x)y}} = \frac{1}{b^{-xy}} = b^{xy}$$

If $x < 0$ and $y < 0$,

$$(b^x)^y = \left(\frac{1}{b^{-x}}\right)^y = \frac{1}{\left(\dfrac{1}{b^{-x}}\right)^{-y}} = \frac{1}{\dfrac{1}{(b^{-x})^{-y}}}$$

$$= \frac{1}{\dfrac{1}{b^{(-x)(-y)}}} = b^{(-x)(-y)} = b^{xy}$$

Thus we see (ii) can be extended.

Example 4

$$(5^2)^{-3} = 5^{-6} \quad \square$$

Example 5

$$[(1.9)^{-2}]^{-4} = (1.9)^8 \quad \square$$

Example 6

$$[(-.4)^5]^{-1} = (-.4)^{-5} \quad \square$$

The justification given for I and II is somewhat long and involved but gives you an understanding of *why* the laws for exponents hold. In order to complete the extensions we need to establish the following two principles:

III. If a and b are nonzero numbers and x is an integer, then

$$(ab)^x = a^x b^x$$

IV. If a and b are nonzero numbers and x is an integer, then

$$\left(\frac{a}{b}\right)^x = \frac{a^x}{b^x}$$

To understand why III holds, notice first that if $x > 0$, the equation holds by (iii).

If $x = 0$, then $(ab)^0 = 1 = 1 \cdot 1 = a^0 \cdot b^0$.
If $x < 0$, then

$$(ab)^x = \frac{1}{(ab)^{-x}}$$

$$= \frac{1}{a^{-x}b^{-x}} \qquad \text{by (iii) since } -x > 0$$

$$= \left(\frac{1}{a^{-x}}\right)\left(\frac{1}{b^{-x}}\right)$$

$$= a^x b^x$$

Thus III holds.

Example 7

$$(3 \cdot 4)^{-5} = (3^{-5})(4^{-5}) \quad \square$$

Example 8

$$\left(\tfrac{2}{3} \cdot \tfrac{4}{7}\right)^{-3} = \left(\tfrac{2}{3}\right)^{-3}\left(\tfrac{4}{7}\right)^{-3} \quad \square$$

To establish IV, we use III and II:

$$\left(\frac{a}{b}\right)^x = \left(a \cdot \frac{1}{b}\right)^x = (a \cdot b^{-1})^x$$

$$= a^x(b^{-1})^x = a^x b^{(-1)x}$$

$$= a^x b^{-x} = a^x\left(\frac{1}{b^x}\right)$$

$$= \frac{a^x}{b^x}$$

Example 9

$$\left(\frac{2}{3}\right)^{-3} = \frac{2^{-3}}{3^{-3}} \quad \square$$

Example 10

$$\left(\frac{-5}{7}\right)^{-4} = \frac{(-5)^{-4}}{7^{-4}} \quad \square$$

We can use all four of these laws and the definitions to simplify expressions.

Example 11

$$\left(\frac{4}{3}\right)^{-2} = \frac{1}{\left(\frac{4}{3}\right)^{2}} = \frac{1}{\left(\frac{4^2}{3^2}\right)} = \frac{3^2}{4^2} = \frac{9}{16}$$

Example 12

$$(2^3 \cdot 3^5 \cdot 7^{-2})(2^{-2} \cdot 3^2 \cdot 5^{-2} \cdot 7^3) = 2^1 \cdot 3^7 \cdot 5^{-2} \cdot 7^1$$

$$= 2^1 \cdot 3^7 \cdot \frac{1}{5^2} \cdot 7^1 = \frac{2 \cdot 3^7 \cdot 7}{5^2}$$

Example 13

$$(a^2 b^3 c^{-2})^3 (a^{-3} b^2 c^3) = a^6 b^9 c^{-6} a^{-3} b^2 c^3 = a^3 b^{11} c^{-3} = \frac{a^3 b^{11}}{c^3}$$

Integral exponents can be used to simplify writing very large or very small positive numbers.

Example 14

$$135,000,000 = (100,000,000)(1.35)$$
$$= 10^8 \cdot 1.35$$

because $10^8 = 100,000,000$. □

Example 15

$$.000476 = (.0001)(4.76)$$
$$= 10^{-4} \cdot 4.76$$

because $10^{-4} = 1/10^4 = .0001$. □

If a number can be expressed as a positive finite decimal, then it can be expressed as the product of an integral power of 10 and a number between 1 and 10. Expressing numbers in this way simplifies computation and eliminates handling many digits. This notation is very useful in scientific and technical computations and is known as *scientific notation*.

Example 16

To express 4,327,000,000 in scientific notation we write it as $10^9 \cdot 4.327$. □

Example 17

To express .0000597 in scientific notation we write it as $10^{-5} \cdot (5.97)$. □

Example 18

To multiply 325,000,000 and 42,000,000,000 we first can rewrite using scientific notation. Thus the product is

$$(10^8 \cdot 3.25)(10^{10} \cdot 4.2) = 10^{18}(3.25 \cdot 4.2)$$
$$= 10^{18}(13.65) = 10^{19}(1.365) \square$$

Example 19

$$\frac{(432,000,000)(7,500,000)}{225,000,000,000} = \frac{10^8(4.32)10^6(7.5)}{10^{11}(2.25)}$$

$$= \frac{10^{14}(4.32)(7.5)}{10^{11}(2.25)}$$

$$= 10^3 \left(\frac{4.32 \cdot 7.5}{2.25} \right)$$

$$= 10^3(14.4) = 10^4(1.44) \quad \text{or} \quad 14{,}400 \square$$

EXERCISES

1. Apply the definitions and the laws of exponents to rewrite the following without using exponents.
   (a) 5^{-2} (b) $(-6)^0$ (c) $(.1)^{-2}$ (d) $(10)^{-3}$
   (e) $4^{-5}4^3$ (f) $(-3)^{-2}$ (g) $(\frac{4}{5})^{-2}$ (h) $(\frac{2}{3})^{-3}$
   (i) $(-\frac{3}{7})^{-4}$ (j) $(.08)^5(.08)^{-6}$

2. Write the following numbers using scientific notation.
   (a) 93,200,000   (b) 758,000,000
   (c) .00004962   (d) .0000000739
   (e) 62,340,000   (f) .4973
   (g) .0008762   (h) .00000000000947

3. The following numbers are expressed in scientific notation. Rewrite them in the usual decimal form.
   (a) $5.47(10)^5$ (b) $6.94(10)^{-3}$
   (c) $8.76(10)^7$ (d) $9.92(10)^{-8}$
   (e) $8.132(10)^{13}$ (f) $6.429(10)^{-5}$
   (g) $3.126(10)^{-4}$ (h) $7.519(10)^{-2}$

4. Use scientific notation in computing the following products and quotients.
   (a) $(84,600,000)(920,000,000,000)$ (b) $(.00926)(.00047)(82,000,000)$
   (c) $(.000761)(.00092)/(.00069)$ (d) $(.00095)^3/(.0021)^4$
   (e) $(.00413)(82,000,000)(.0052)$ (f) $(15,000,000)/(225,000)$
   (g) $(.082)^3/(250)^4$ (h) $(90,900)/(300)^3$

*5. Let $D = \{p(10)^n : p \in Z \text{ and } n \in Z\}$. Show that if $x \in D$ and $y \in D$, then $x + y \in D$ and $xy \in D$. Does the set D have a familiar name from arithmetic? [*Hint:* Write $241(10)^{-2}$ without using a negative exponent or a fraction line.]

2. Square Roots

In Chapter 2 we considered equations in one variable or several variables but in no case did a variable appear in a denominator or was a variable multiplied by itself or another variable. The equations considered in Chapter 2 could all be solved using only addition, subtraction, multiplication, and division. Our knowledge of exponents now enables us to write equations which cannot be solved by means of these arithmetic operations.

Example 1

Solve the equation

$$x^2 = 4$$

We know that 2 and -2 are solutions to this equation but we cannot obtain the equations $x = 2$ or $x = -2$ by "doing the same thing to both sides" of $x^2 = 4$. Factoring and reasoning, however, enable us to determine that these are two solutions.

If

$$x^2 = 4$$
$$x^2 - 4 = 0$$

and therefore

$$(x - 2)(x + 2) = 0$$

The product of two numbers is 0 only if at least one of them is 0. Thus if

$$(x - 2)(x + 2) = 0$$

either

$$x - 2 = 0$$

or

$$x + 2 = 0$$

If $x - 2 = 0$, then $x = 2$. If $x + 2 = 0$, then $x = -2$. Thus if $x^2 = 4$, then $x = 2$ or $x = -2$. □

The argument given in Example 1 generalizes as follows:

If $a \in R$ and $x^2 = a^2$, then $x^2 - a^2 = 0$. Factoring yields the equation $(x - a)(x + a) = 0$ and either $x - a = 0$ or $x + a = 0$. Therefore $x = a$ or $x = -a$. Thus if $x^2 = a^2$, either $x = a$ or $x = -a$. If $a \neq 0$, the equation $x^2 = a^2$ has two solutions; if $a = 0$, there is only one solution.

Example 2

The solutions of $x^2 = 9$ are 3 and -3. □

Example 3

The solutions of $y^2 = \dfrac{4}{25}$ are $\dfrac{2}{5}$ and $-\dfrac{2}{5}$. $\square$

Example 4

The solutions of $x^2 = .01$ are $.1$ and $-.1$. $\square$

What about an equation such as $x^2 = 2$? We have no known way of factoring 2 as the square of a rational number. In fact, no such rational number exists. We will not give a proof here but will delay the explanation until Section 4.

We would expect equations such as $x^2 = 2$ to have solutions in R even though no solution exists in Q. We can imagine a square with area 2 and its side must be a number whose square is 2. Axiom 12 can be used to show that such a number exists and, in fact, *if r is any positive number, the equation $x^2 = r$ can be shown to have exactly one positive solution and one negative solution.* The proof is long and difficult and beyond the scope and intent of this text. We will limit ourselves to discussing expressions for these numbers and the corresponding arithmetic.

If $r \geq 0$, then $\sqrt{r}$ denotes the nonnegative number whose square is r. If $r \geq 0$, $\sqrt{r}$ is called the *square root* of r or the *principle square root* of r.

Example 5

$(\sqrt{3})^2 = 3$ $\square$

Example 6

$(\sqrt{11})^2 = 11$ $\square$

Example 7

$\sqrt{9} = 3$ since $3 > 0$ and $3^2 = 9$. $\square$

If we square a negative number and then take the square root, we obtain a positive number. Thus if $y \in R$, then $\sqrt{y^2} = |y|$ because $\sqrt{y^2}$ must be nonnegative. This is why *the root-taking process is not quite the reverse of taking powers.*

Example 8

$\sqrt{(-2)^2} = 2$ $\square$

Example 9

$\sqrt{(-.4)^2} = .4$ $\square$

An expression such as $\sqrt{a}$ is called a *radical*, the symbol $\sqrt{}$ is called a *radical sign*, and "a" is called the *radicand*. As with exponents, these symbols are just bits of notation and computational techniques involving them are based on their meaning.

If a and b are nonnegative numbers, can we find a simple expression for the product of their square roots; that is, can we simplify $\sqrt{a}\sqrt{b}$?

Example 10

$$
\begin{aligned}
[(\sqrt{2})(\sqrt{3})]^2 &= [(\sqrt{2})(\sqrt{3})][(\sqrt{2})(\sqrt{3})] \\
&= [(\sqrt{2})(\sqrt{2})][(\sqrt{3})(\sqrt{3})] \\
&= 2 \cdot 3 = 6
\end{aligned}
$$

Thus $(\sqrt{3})(\sqrt{2})$ is a nonnegative number (since $\sqrt{2}$ and $\sqrt{3}$ are nonnegative) whose square is 6. Since there is only one nonnegative number whose square is 6, it follows that

$$(\sqrt{2})(\sqrt{3}) = \sqrt{6} \;\square$$

The argument given in Example 10 can be generalized as follows:

If a and b are nonnegative numbers, then

$$
\begin{aligned}
[\sqrt{a}\sqrt{b}]^2 &= [\sqrt{a}\sqrt{b}][\sqrt{a}\sqrt{b}] \\
&= [\sqrt{a}\sqrt{a}][\sqrt{b}\sqrt{b}] \\
&= [\sqrt{a}]^2[\sqrt{b}]^2 \\
&= ab
\end{aligned}
$$

Since $\sqrt{a} \geq 0$ and $\sqrt{b} \geq 0$, $\sqrt{a}\sqrt{b} \geq 0$. Thus $\sqrt{a}\sqrt{b}$ is a nonnegative number whose square is ab. Since only one such number exists, $\sqrt{a}\sqrt{b}$ must be $\sqrt{ab}$. Thus

$$\text{if} \quad a \geq 0 \quad \text{and} \quad b \geq 0, \quad \sqrt{a}\sqrt{b} = \sqrt{ab}$$

Example 11

$$\sqrt{5}\sqrt{3} = \sqrt{15} \;\square$$

Example 12

$$\sqrt{6}\sqrt{2} = \sqrt{12} \;\square$$

Example 13

$$\sqrt{5}\sqrt{7} = \sqrt{35} \;\square$$

The principle just established can be used to simplify expressions involving radicals.

Example 14

$$\sqrt{12} = \sqrt{4 \cdot 3} = \sqrt{4}\sqrt{3} = 2\sqrt{3} \ \square$$

Example 15

$$\sqrt{45} = \sqrt{9 \cdot 5} = \sqrt{9}\sqrt{5} = 3\sqrt{5} \ \square$$

Example 16

$$\sqrt{16x^5} = \sqrt{16x^4 \cdot x} = \sqrt{16x^4}\sqrt{x} = 4x^2\sqrt{x} \ \square$$

Example 17

$$\sqrt{y^2 x} = \sqrt{y^2}\sqrt{x} = |y|\sqrt{x} \ \square$$

In Example 17 we needed the absolute value symbol because y can be positive, negative, or zero but, in any case, $\sqrt{y^2} = |y|$.

The basic technique for simplifying radicals with constant or polynomial radicands is as follows:

A. Write the radicand as a product of a perfect square and another expression which does not have a perfect square as a factor.
B. Factor the expression into the product of two radicals.
C. Simplify the expression with the perfect square radicand.

Example 18

$$\sqrt{20x^3 y^5} = \sqrt{4x^2 y^4 \cdot 5xy} = \sqrt{4x^2 y^4}\sqrt{5xy}$$
$$= |2xy^2|\sqrt{5xy} \ \square$$

Example 19

$$\sqrt{98} = \sqrt{49 \cdot 2} = \sqrt{49}\sqrt{2} = 7\sqrt{2} \ \square$$

Example 20

$$\sqrt{.4} = \sqrt{(.04)10} = \sqrt{.04}\sqrt{10} = .2\sqrt{10} \ \square$$

Example 21

$$\sqrt{75(x - y)^5} = \sqrt{25(x - y)^4 \cdot 3(x - y)}$$
$$= \sqrt{25(x - y)^4}\sqrt{3(x - y)}$$
$$= 5(x - y)^2\sqrt{3(x - y)}$$

Notice that no absolute value symbol is needed because $5(x - y)^2 \geq 0$ for all values of x and y. $\square$

Example 22

By checking the definition, we see that $(\sqrt{2}/\sqrt{3})^2 = 2/3$, that is,

$$\left(\frac{\sqrt{2}}{\sqrt{3}}\right)^2 = \frac{(\sqrt{2})^2}{(\sqrt{3})^2} = \frac{2}{3}$$

and since $\sqrt{2} > 0$ and $\sqrt{3} > 0$, $\sqrt{2}/\sqrt{3} > 0$. Thus $\sqrt{2}/\sqrt{3}$ is the non-negative number whose square is 2/3 and, therefore

$$\frac{\sqrt{2}}{\sqrt{3}} = \sqrt{\frac{2}{3}} \quad \square$$

The argument above generalizes as follows:

If $a \geq 0$ and $b > 0$, then

$$\left(\frac{\sqrt{a}}{\sqrt{b}}\right)^2 = \frac{(\sqrt{a})^2}{(\sqrt{b})^2} = \frac{a}{b}$$

Since $\sqrt{a} \geq 0$ and $\sqrt{b} > 0$, $\sqrt{a}/\sqrt{b} \geq 0$.

Thus $\sqrt{a}/\sqrt{b}$ is the nonnegative number whose square is a/b and, therefore

$$\frac{\sqrt{a}}{\sqrt{b}} = \sqrt{\frac{a}{b}}$$

Example 23

$$\frac{\sqrt{15}}{\sqrt{5}} = \sqrt{\frac{15}{5}} = \sqrt{3} \quad \square$$

Example 24

$$\frac{\sqrt{10}}{\sqrt{.1}} = \sqrt{\frac{10}{.1}} = \sqrt{100} = 10 \quad \square$$

This principle, along with the previously established techniques, can be used to simplify expressions having a radicand in fractional form. If the denominator of the radicand is a perfect square, we can apply this principle to obtain a fraction in which the denominator contains no radicals.

Example 25

$$\sqrt{\frac{5}{9}} = \frac{\sqrt{5}}{\sqrt{9}} = \frac{\sqrt{5}}{3} \quad \square$$

Example 26

$$\sqrt{\frac{7}{4}} = \frac{\sqrt{7}}{\sqrt{4}} = \frac{\sqrt{7}}{2} \ \square$$

Example 27

$$\sqrt{\frac{5y^3}{x^4}} = \frac{\sqrt{5y^3}}{\sqrt{x^4}} = \frac{y\sqrt{5y}}{x^2} \ \square$$

If we have a fractional radicand in which the denominator is not a perfect square, we can multiply the numerator and denominator by a suitable expression so as to make the denominator a perfect square and then simplify.

Example 28

$$\sqrt{\frac{2}{3}} = \sqrt{\frac{2 \cdot 3}{3 \cdot 3}} = \sqrt{\frac{6}{9}} = \frac{\sqrt{6}}{\sqrt{9}} = \frac{\sqrt{6}}{3} \ \square$$

Example 29

$$\sqrt{\frac{5}{8}} = \sqrt{\frac{10}{16}} = \frac{\sqrt{10}}{\sqrt{16}} = \frac{\sqrt{10}}{4} \ \square$$

Example 30

$$\sqrt{\frac{3zy^3}{2x^3}} = \sqrt{\frac{6zy^3x}{4x^4}} = \frac{\sqrt{6xy^3x}}{\sqrt{4x^4}} = \frac{\sqrt{6zy^3x}}{2x^2} = \frac{|y|\sqrt{6zyx}}{2x^2} \ \square$$

Such expressions can also be simplified by rewriting as a quotient of radicals and then multiplying the numerator and denominator by some expression which will yield a denominator containing no radicals.

Example 31

$$\sqrt{\frac{5}{6}} = \frac{\sqrt{5}}{\sqrt{6}} = \frac{\sqrt{5}\sqrt{6}}{\sqrt{6}\sqrt{6}} = \frac{\sqrt{30}}{6} \ \square$$

Example 32

$$\sqrt{\frac{3}{8}} = \frac{\sqrt{3}}{\sqrt{8}} = \frac{\sqrt{3}\sqrt{2}}{\sqrt{8}\sqrt{2}} = \frac{\sqrt{6}}{\sqrt{16}} = \frac{\sqrt{6}}{4} \ \square$$

Example 33

$$\sqrt{\frac{2x}{3y^3}} = \frac{\sqrt{2x}}{\sqrt{3y^3}} = \frac{\sqrt{2x}\sqrt{3y}}{\sqrt{3y^3}\sqrt{3y}} = \frac{\sqrt{6xy}}{\sqrt{9y^4}} = \frac{\sqrt{6xy}}{3y^2} \quad \square$$

Either method is satisfactory and the reader should be able to use both equally well.

The special product

$$(a + b)(a - b) = a^2 - b^2$$

can be used to simplify fractions in which the denominator is the sum or difference of square roots. If it is the sum, we multiply the numerator and denominator by the difference and vice versa.

Example 34

$$\frac{5}{\sqrt{3} - \sqrt{2}} = \frac{5(\sqrt{3} + \sqrt{2})}{(\sqrt{3} - \sqrt{2})(\sqrt{3} + \sqrt{2})} = \frac{5(\sqrt{3} + \sqrt{2})}{3 - 2}$$

$$= \frac{5(\sqrt{3} + \sqrt{2})}{1} = 5(\sqrt{3} + \sqrt{2}) \quad \square$$

Example 35

$$\frac{\sqrt{5} - \sqrt{3}}{\sqrt{5} + \sqrt{3}} = \frac{(\sqrt{5} - \sqrt{3})(\sqrt{5} - \sqrt{3})}{(\sqrt{5} + \sqrt{3})(\sqrt{5} - \sqrt{3})}$$

$$= \frac{5 - 2\sqrt{3}\sqrt{5} + 3}{5 - 3}$$

$$= \frac{8 - 2\sqrt{15}}{2}$$

$$= 4 - \sqrt{15} \quad \square$$

Example 36

$$\frac{4}{\sqrt{11} - \sqrt{5}} = \frac{4(\sqrt{11} + \sqrt{5})}{(\sqrt{11} - \sqrt{5})(\sqrt{11} + \sqrt{5})}$$

$$= \frac{4(\sqrt{11} + \sqrt{5})}{11 - 5}$$

$$= \frac{4(\sqrt{11} + \sqrt{5})}{6}$$

$$= (\tfrac{2}{3})(\sqrt{11} + \sqrt{5})$$

The basic idea is to multiply both the numerator and denominator by a suitable number so that the resulting denominator contains no radicals. We will find this technique very useful when we wish to approximate such numbers by finite decimals in Section 4.

EXERCISES

1. Rewrite the following using only one radical.
   (a) $\sqrt{7}\sqrt{5}$
   (b) $\sqrt{19}\sqrt{8}$
   (c) $\sqrt{4.3}\sqrt{9.2}$
   (d) $\sqrt{5xy^2}\sqrt{11x^3y^4}$
   (e) $\sqrt{5/7}\sqrt{3/5}$
   (f) $\sqrt{.5}\sqrt{10}$
   (g) $\sqrt{2/3}\sqrt{243/7}$
   (h) $\sqrt{5x^2y^3/6z^5}\sqrt{2xy^5/7z^3}$

2. Rewrite the following using only one radical.
   (a) $\sqrt{11}\sqrt{13}$
   (b) $\sqrt{14}\sqrt{6}$
   (c) $\sqrt{8.7}\sqrt{1.2}$
   (d) $\sqrt{6x^5y^7}\sqrt{2xy^9}$
   (e) $\sqrt{4/11}\sqrt{7/4}$
   (f) $\sqrt{2/9}\sqrt{135/4}$
   (g) $\sqrt{.6}\sqrt{.8}$
   (h) $\sqrt{\dfrac{7x^5y^7}{11z^5}}\sqrt{\dfrac{33z^3x}{35y^3}}$

3. Simplify as in Examples 14–17.
   (a) $\sqrt{48}$
   (b) $\sqrt{50}$
   (c) $\sqrt{27}$
   (d) $\sqrt{63}$
   (e) $\sqrt{8x^5y^4}$
   (f) $\sqrt{32a^9b^6c^8}$
   (g) $\sqrt{242(x+2y)^7}$
   (h) $\sqrt{68x^7y^8z^9}$
   (i) $\sqrt{(-3)^2x^2y^6}$
   (j) $\sqrt{(-5)^4x^2y^4}$

4. Simplify as in Examples 14–17.
   (a) $\sqrt{28}$
   (b) $\sqrt{108}$
   (c) $\sqrt{75}$
   (d) $\sqrt{52}$
   (e) $\sqrt{20x^7y^4}$
   (f) $\sqrt{72x^5y^9}$
   (g) $\sqrt{45a^4b^6c^9}$
   (h) $\sqrt{242x^{17}y^{23}}$
   (i) $\sqrt{(-7)^2a^2b^6}$
   (j) $\sqrt{(-2)^4a^2b^4}$

5. Simplify. Your answer should not have a radical in the denominator.
   (a) $\sqrt{7/9}$
   (b) $\sqrt{11/49}$
   (c) $\sqrt{3/5}$
   (d) $\sqrt{25x^2y^3/49z^5}$
   (e) $\sqrt{16/27}$
   (f) $\sqrt{5/17}$
   (g) $\sqrt{x/3y}$
   (h) $\sqrt{17x/50y^3}$
   (i) $\sqrt{7}/\sqrt{3}$
   (j) $\sqrt{2}/\sqrt{5}$
   (k) $\sqrt{18}/\sqrt{2}$
   (l) $\sqrt{a}/\sqrt{b}$

6. Simplify. Your answer should not have a radical in the denominator.
   (a) $\sqrt{5/16}$
   (b) $\sqrt{25/36}$
   (c) $\sqrt{13/72}$
   (d) $\sqrt{3/7}$
   (e) $\sqrt{8/121}$
   (f) $\sqrt{3/19}$
   (g) $\sqrt{a/7b}$
   (h) $\sqrt{13a/18b^3}$
   (i) $\sqrt{3}/\sqrt{11}$
   (j) $\sqrt{5}/\sqrt{2}$
   (k) $\sqrt{12}/\sqrt{3}$
   (l) $\sqrt{x}/\sqrt{y}$

7. Simplify your answers to Exercise 1 and Exercise 2.

8. All of your answers to Exercises 3–7 could be written with the absolute value symbol. Which ones *must* include absolute value signs? Justify your answer.

9. Express each of the following with no radicals in the denominator. Simplify as much as you can.

(a) $\dfrac{\sqrt{5} - \sqrt{2}}{\sqrt{5} + \sqrt{2}}$ (b) $\dfrac{2\sqrt{7} + 5\sqrt{3}}{\sqrt{7} - \sqrt{3}}$

(c) $\dfrac{\sqrt{13} + 4\sqrt{7}}{\sqrt{13} - \sqrt{7}}$ (d) $\dfrac{7\sqrt{3} - 2\sqrt{5}}{\sqrt{3} + \sqrt{5}}$

(e) $\dfrac{8\sqrt{7} - 3\sqrt{5}}{\sqrt{7} + \sqrt{5}}$ (f) $\dfrac{p\sqrt{t} - q\sqrt{s}}{\sqrt{t} + \sqrt{s}}$

(g) $\dfrac{p\sqrt{t} - q\sqrt{s}}{\sqrt{t} - \sqrt{s}}$ (h) $\dfrac{1}{\sqrt{a} - \sqrt{b}}$

(i) $\dfrac{1}{\sqrt{a} + \sqrt{b}}$ (j) $\dfrac{\sqrt{a} - \sqrt{b}}{\sqrt{a} + \sqrt{b}}$

10. Express each of the following with no radicals in the denominator. Simplify as much as you can.

(a) $\dfrac{\sqrt{11} - \sqrt{7}}{\sqrt{11} + \sqrt{7}}$ (b) $\dfrac{2\sqrt{7} - 3\sqrt{5}}{\sqrt{7} + \sqrt{5}}$

(c) $\dfrac{\sqrt{17} + 5\sqrt{13}}{\sqrt{17} - \sqrt{13}}$ (d) $\dfrac{5\sqrt{6} + 2\sqrt{3}}{\sqrt{6} - \sqrt{3}}$

(e) $\dfrac{11\sqrt{19} + 2\sqrt{7}}{\sqrt{19} - \sqrt{7}}$ (f) $\dfrac{x\sqrt{a} + y\sqrt{b}}{\sqrt{a} - \sqrt{b}}$

(g) $\dfrac{x\sqrt{a} + y\sqrt{b}}{\sqrt{a} + \sqrt{b}}$ (h) $\dfrac{1}{\sqrt{x} + \sqrt{y}}$

(i) $\dfrac{1}{\sqrt{x} - \sqrt{y}}$ (j) $\dfrac{\sqrt{x} + \sqrt{y}}{\sqrt{x} - \sqrt{y}}$

11. By direct multiplication and simplification, show that

$$\frac{B + \sqrt{B^2 - 4AC}}{2A} \cdot \frac{B - \sqrt{B^2 - 4AC}}{2A} = \frac{C}{A}$$

3. Other Roots and Radicals

The square root of a number a was defined as the nonnegative solution to the equation $x^2 = a$ if $a \geq 0$. The equation $x^3 = 8$ has only one solution, namely 2, and the equation $x^3 = -27$ has -3 as its only solution. The equation $x^4 = 16$ has both 2 and -2 as solutions. In general, if n is an *odd* positive integer and *a* is any real number, then the equation $x^n = a$ has exactly *one* solution. The solution is *positive if a is positive* and *negative if a is negative*. If *n* is an *even* positive integer and $a > 0$, the equation $x^n = a$ has *two* solutions, one positive and one negative. (These facts can be proved using Axiom 12 but the proofs are very difficult and to include them would not add to the student's understanding of the concepts but would tend to bog him down with technical fine points. We will not include them in this text.)

With these facts in mind, we now make the following definition:

> If *n* is an odd positive integer and *r* is a real number, then $\sqrt[n]{r}$ is the unique solution of the equation $x^n = r$. If *n* is an even positive integer, and $r \geq 0$, $\sqrt[n]{r}$ is the unique *nonnegative* solution of the equation $x^n = r$.

The expression $\sqrt[n]{r}$ is called "the *n*th root of *r*" (sometimes the principal *n*th root of *r*). The number $\sqrt[3]{r}$ is frequently called the *cube root* of *r*. As with square roots, such expressions are called *radicals*. In the expression $\sqrt[n]{r}$, *n* is called the *index* and *r* the *radicand*.

Notice that just as $\sqrt{-5}$ did not exist, $\sqrt[n]{r}$ does not exist when *n* is even and $r < 0$ because $x^n \geq 0$ for all real numbers *x* if *n* is even.

In dealing with square roots we found that

$$\sqrt{a}\sqrt{b} = \sqrt{ab}$$

if $\sqrt{a}$ and $\sqrt{b}$ both exist, that is, if $a \geq 0$ and $b \geq 0$.

Suppose *a* and *b* are real numbers and consider the product $\sqrt[3]{a}\sqrt[3]{b}$. Using the third law of exponents we have

$$(\sqrt[3]{a}\sqrt[3]{b})^3 = (\sqrt[3]{a})^3(\sqrt[3]{b})^3 = ab$$

Thus $\sqrt[3]{a}\sqrt[3]{b}$ is a solution of the equation $x^3 = ab$. Since only one solution exists, namely $\sqrt[3]{ab}$, we must have $\sqrt[3]{a}\sqrt[3]{b} = \sqrt[3]{ab}$.

Example 1

$$\sqrt[3]{2}\sqrt[3]{5} = \sqrt[3]{10} \quad \square$$

Example 2

$$\sqrt[3]{7}\sqrt[3]{-5} = \sqrt[3]{-35} \quad \square$$

Example 3

$$\sqrt[3]{4}\sqrt[3]{2} = \sqrt[3]{8} = 2 \quad \square$$

Suppose now that $a \geq 0$ and $b \geq 0$. Then $\sqrt[4]{a}$ and $\sqrt[4]{b}$ exist and, by the same law of exponents,

$$(\sqrt[4]{a}\sqrt[4]{b})^4 = (\sqrt[4]{a})^4(\sqrt[4]{b})^4 = ab$$

Thus $\sqrt[4]{a}\sqrt[4]{b}$ is a solution of $x^4 = ab$. Since $\sqrt[4]{a} \geq 0$ and $\sqrt[4]{b} \geq 0$, $\sqrt[4]{a}\sqrt[4]{b} \geq 0$ and, therefore $\sqrt[4]{a}\sqrt[4]{b} = \sqrt[4]{ab}$.

Example 4

$$\sqrt[4]{3}\sqrt[4]{5} = \sqrt[4]{15} \ \square$$

Example 5

$$\sqrt[4]{xy}\sqrt[4]{xy^2} = \sqrt[4]{x^2y^3} \ \square$$

Example 6

$$\sqrt[4]{3x^3y}\sqrt[4]{5xyz^3} = \sqrt[4]{15x^3y^2z^3} \ \square$$

In general, if $\sqrt[n]{a}$ and $\sqrt[n]{b}$ both exist, then

$$\sqrt[n]{a}\sqrt[n]{b} = \sqrt[n]{ab}$$

The explanation is essentially the same as the cases above and we leave it as an exercise. The student must remember that the equation only holds if all expressions have meaning. A sentence such as

$$\sqrt[4]{-3}\sqrt[4]{-5} = \sqrt[4]{15}$$

is nonsense because $\sqrt[4]{-3}$ and $\sqrt[4]{-5}$ do not exist.

Using reasoning along the same line as we followed with square roots and with the products above, it is easy to show that if $\sqrt[n]{a}$ and $\sqrt[n]{b}$ both exist and $\sqrt[n]{b} \neq 0$, then

$$\frac{\sqrt[n]{a}}{\sqrt[n]{b}} = \sqrt[n]{\frac{a}{b}}$$

We leave the explanations as exercises.

Since our simplification techniques for square roots were all based on the two laws

$$\sqrt{a}\sqrt{b} = \sqrt{ab} \quad \text{and} \quad \frac{\sqrt{a}}{\sqrt{b}} = \sqrt{\frac{a}{b}}$$

we can develop similar simplification techniques for nth roots.

Example 7

$$\sqrt[3]{24} = \sqrt[3]{8 \cdot 3} = \sqrt[3]{8}\sqrt[3]{3} = 2\sqrt[3]{3} \ \square$$

Example 8

$$\sqrt[4]{162} = \sqrt[4]{81 \cdot 2} = \sqrt[4]{81}\sqrt[4]{2} = 3\sqrt[4]{2} \ \square$$

Example 9

$$\sqrt[3]{-5} = \sqrt[3]{(-1) \cdot 5} = \sqrt[3]{-1}\sqrt[3]{5} = (-1)\sqrt[3]{5} = -\sqrt[3]{5} \ \square$$

Example 10

$$\sqrt[7]{-a^2} = \sqrt[7]{(-1)a^2} = \sqrt[7]{-1}\sqrt[7]{a^2} = -1\sqrt[7]{a^2} = -\sqrt[7]{a^2} \ \square$$

From the last two examples we see that an odd nth root of a negative number can always be rewritten as the negative of an nth root of a positive number. That is, if $a < 0$ and n is odd, then

$$\sqrt[n]{a} = \sqrt[n]{(-1)|a|} = \sqrt[n]{-1}\sqrt[n]{|a|} = -1\sqrt[n]{|a|} = -\sqrt[n]{|a|}$$

When variables appear in the radicand we must make sure that the rewritten form takes into account the possibilities for the variable being positive or negative.

$$\sqrt[4]{x^5 y^3} = \sqrt[4]{x^4 x y^3} = \sqrt[4]{x^4}\sqrt[4]{xy^3} = |x|\sqrt[4]{xy^3}$$

In this case, if $x < 0$ and $y < 0$, $\sqrt[4]{x^5 y^3}$ exists and is positive whereas $x\sqrt{xy^3}$ is negative. We needed the absolute value sign to indicate that we had a positive number.

The technique for simplifying products is essentially the same as we used for square roots. The only difference is that we need to factor the radicand into the product of an nth power and another factor containing no nth powers.

Example 11

$$\sqrt[4]{\frac{3}{2}} = \sqrt[4]{\frac{24}{16}} = \frac{\sqrt[4]{24}}{\sqrt[4]{16}} = \frac{\sqrt[4]{24}}{2} = \frac{1}{2}\sqrt[4]{24} \ \square$$

Example 12

$$\sqrt[3]{\frac{3}{5}} = \frac{\sqrt[3]{3}}{\sqrt[3]{5}} = \frac{\sqrt[3]{3}\sqrt[3]{25}}{\sqrt[3]{5}\sqrt[3]{25}} = \frac{\sqrt[3]{75}}{\sqrt[3]{125}} = \frac{\sqrt[3]{75}}{5} \ \square$$

Example 13

$$\sqrt[3]{.1} = \sqrt[3]{\frac{1}{10}} = \sqrt[3]{\frac{100}{1000}} = \frac{1}{10}\sqrt[3]{100} = .1\sqrt[3]{100} \ \square$$

Example 14

$$\sqrt[5]{\frac{x}{y^3}} = \sqrt[5]{\frac{xy^2}{y^5}} = \frac{\sqrt[5]{xy^2}}{\sqrt[5]{y^5}} = \frac{\sqrt[5]{xy^2}}{y} \ \square$$

Expressions such as

$$\frac{6}{\sqrt[3]{5} - \sqrt[3]{2}}$$

can be simplified but not as easily as when all the roots were square roots. We can use the special product $(x - y)(x^2 + xy + y^2) = x^3 - y^3$ to simplify the expression.

Example 15

$$\frac{6}{\sqrt[3]{5} - \sqrt[3]{2}} = \frac{6[(\sqrt[3]{5})^2 + \sqrt[3]{5}\sqrt[3]{2} + (\sqrt[3]{2})^2]}{(\sqrt[3]{5} - \sqrt[3]{2})[(\sqrt[3]{5})^2 + \sqrt[3]{5}\sqrt[3]{2} + (\sqrt[3]{2})^2]}$$

$$= \frac{6[\sqrt[3]{25} + \sqrt[3]{10} + \sqrt[3]{4}]}{(\sqrt[3]{5})^3 - (\sqrt[3]{2})^3}$$

$$= \frac{6[\sqrt[3]{25} + \sqrt[3]{10} + \sqrt[3]{4}]}{3} = 2[\sqrt[3]{25} + \sqrt[3]{10} + \sqrt[3]{4}] \quad \square$$

We will see in Section 4 that this form is more useful for determining decimal approximations.

EXERCISES

1. Rewrite using only one radical.
   (a) $\sqrt[3]{4}\sqrt[3]{23}$
   (b) $\sqrt[4]{4}\sqrt[4]{12}$
   (c) $\sqrt[5]{12}\sqrt[5]{40}$
   (d) $\sqrt[3]{.1}\sqrt[3]{.08}$
   (e) $\sqrt[3]{x^2y^4}\sqrt[3]{4xy^7}$
   (f) $\sqrt[5]{6x^4y^3z^3}\sqrt[5]{48x^3y^4z^9}$
   (g) $\sqrt[3]{\frac{2}{3}}\sqrt[3]{\frac{5}{9}}$
   (h) $\sqrt[5]{\frac{5}{4}}\sqrt[5]{\frac{9}{16}}$
   (i) $\sqrt[3]{9x^2y^5}\sqrt[3]{18x^2y^2}$
   (j) $\sqrt[3]{8x^5y^6}\sqrt[7]{16x^3y^5}$

2. Rewrite using only one radical.
   (a) $\sqrt[3]{5}\sqrt[3]{11}$
   (b) $\sqrt[4]{6}\sqrt[4]{24}$
   (c) $\sqrt[5]{20}\sqrt[5]{48}$
   (d) $\sqrt[3]{.18}\sqrt[3]{.3}$
   (e) $\sqrt[3]{a^2b^4}\sqrt[3]{4ab^7}$
   (f) $\sqrt[5]{12a^5b^2c^4}\sqrt[5]{48a^3b^4c^9}$
   (g) $\sqrt[3]{\frac{7}{3}}\sqrt[3]{\frac{2}{9}}$
   (h) $\sqrt[5]{\frac{7}{8}}\sqrt[5]{\frac{9}{20}}$
   (i) $\sqrt[3]{18a^2b^4}\sqrt[3]{15a^2b^5}$
   (j) $\sqrt[7]{16a^5y^5}\sqrt[7]{20a^3y^4}$

3. Simplify.
   (a) $\sqrt[3]{54}$
   (b) $\sqrt[4]{48x^6y^{10}}$
   (c) $\sqrt[5]{.000001}$
   (d) $\sqrt[3]{-16}$
   (e) $\sqrt[3]{-17}$
   (f) $\sqrt[3]{5/4}$
   (g) $\sqrt[4]{5/9}$
   (h) $\sqrt[3]{5/x}$
   (i) $\sqrt[7]{256x^{39}y^{47}z^{11}}$
   (j) $\sqrt[3]{-54x^7y^{11}z}$
   (k) $\sqrt[3]{9}/\sqrt[3]{6}$
   (l) $\sqrt[5]{4}/\sqrt[5]{8}$

4. Simplify.

(a) $\sqrt[3]{40}$ (b) $\sqrt[4]{80a^6b^9}$ (c) $\sqrt[5]{.00000001}$

(d) $\sqrt[3]{-54}$ (e) $\sqrt[3]{-23}$ (f) $\sqrt[3]{7/9}$

(g) $\sqrt[4]{5/4}$ (h) $\sqrt[3]{7/b}$ (i) $\sqrt{1024a^{37}b^{45}c^{17}}$

(j) $\sqrt[3]{-16a^8b^{10}c}$ (k) $\sqrt[3]{4}/\sqrt[3]{6}$ (l) $\sqrt[5]{9}/\sqrt[5]{54}$

5. Simplify your answers to Exercises 1 and 2.
6. Which of your answers to Exercises 3–5 *must* use absolute value symbols? Justify your answer.
7. Rewrite the following with no radicals appearing in the denominator. Simplify as much as possible.

(a) $\dfrac{1}{\sqrt[3]{7} - \sqrt[3]{2}}$ (b) $\dfrac{14}{\sqrt[3]{5} + \sqrt[3]{2}}$

(c) $\dfrac{1}{\sqrt[3]{a} - \sqrt[3]{b}}$ (d) $\dfrac{1}{\sqrt[3]{a} + \sqrt[3]{b}}$

8. Rewrite the following with no radicals appearing in the denominator.

(a) $\dfrac{1}{\sqrt[3]{5} - \sqrt[3]{3}}$ (b) $\dfrac{15}{\sqrt[3]{7} + \sqrt[3]{4}}$

(c) $\dfrac{1}{\sqrt[3]{x} + \sqrt[3]{y}}$ (d) $\dfrac{1}{\sqrt[3]{x} - \sqrt[3]{y}}$

9. Explain why $\sqrt[3]{a}/\sqrt[3]{b} = \sqrt[3]{a/b}$ if $b \neq 0$.
10. Explain why $\sqrt[4]{a}/\sqrt[4]{b} = \sqrt[4]{a/b}$ if $a \geq 0$ and $b \geq 0$.
*11. Explain why $\sqrt[n]{a}\sqrt[n]{b} = \sqrt[n]{ab}$ if n is a positive integer and $\sqrt[n]{a}$ and $\sqrt[n]{b}$ both exist.
*12. Explain why $\sqrt[n]{a}/\sqrt[n]{b} = \sqrt[n]{a/b}$ if n is a positive integer, $\sqrt[n]{a}$ exists, and $\sqrt[n]{b}$ exists and is not zero.

4. Decimal Approximations of Numbers Expressed Using Radicals

In applications and practical problems we would usually not be content with answers such as $5\sqrt{2}$ or $\sqrt[3]{93}$. Although these are perfectly good expressions for numbers, they do not give us the same quantitative grasp or feeling as do 7.070 and 4.531. We like to be able to compare numbers and numbers expressed in decimal form are the easiest to compare. In this section we will be concerned with decimal *approximations* for square roots and cube roots.

In Sections 2 and 3 we attempted to simplify radicals by suitable factoring of the radicand and then splitting the root into two products, one of which could be rewritten without a radical. Could more simplification take place? For example, might there be a rational number a/b such that $a/b = \sqrt{3}$? Might we be able to carry our simplification further? Consider, in particular, $\sqrt{2}$. Can there exist integers a and b such that $a/b = \sqrt{2}$, that is, $(a/b)^2 = 2$? The following indirect argument shows that $\sqrt{2}$ is not in Q, the set of rational numbers.

If $\sqrt{2}$ is rational, then there are integers p and q such that $p/q = \sqrt{2}$ and furthermore, we can assume that p/q is in lowest terms. (If it is not, we can reduce it to an expression that is in lowest terms and proceed.) Since $p/q = \sqrt{2}$,

$$\left(\frac{p}{q}\right)^2 = (\sqrt{2})^2$$

Thus

$$\frac{p^2}{q^2} = 2$$

and furthermore,

$p^2 = 2q^2$ (Multiplying both sides by q^2)

Since $p^2 = 2q^2$, p^2 must be even. If p^2 is even, p must be even. (You may have to think about that for a minute.) Thus for some integer k, $p = 2k$. Substituting we have

$(2k)^2 = 2q^2$

Thus $2 \cdot 2 \cdot k^2 = 2q^2$ and, therefore

$2k^2 = q^2$ (Dividing both sides by 2)

Then q^2 also is even, and therefore q must be even. Since p is even and q is even, both are divisible by 2, and p/q is not in lowest terms. But this contradicts our assumption and therefore $\sqrt{2}$ cannot be rational.

The following more general result can be shown but we will not give a proof in this text.

If n is a positive integer greater than 1 and a is a nonnegative integer, then $\sqrt[n]{a}$ is rational if and only if $\sqrt[n]{a}$ is an integer.

Thus, for example, since $2 < \sqrt{6} < 3$, $\sqrt{6}$ is not an integer, and therefore is irrational. If a is an integer such that $\sqrt{a}$ is an integer, then a is said to be a *perfect square*. The numbers 4, 9, 16, 25, 49, and 100 are examples of perfect squares.

Although we cannot express numbers such as $\sqrt{2}$ or $\sqrt[4]{5}$ by means of fractions, we can *approximate* them by fractions and, in fact, we can approximate as closely as desired by finite decimals.

Example 1

1.414 is a finite decimal approximation for $\sqrt{2}$ because

$(1.414)^2 = 1.999396$ □

Example 2

1.732 is a finite decimal approximation for $\sqrt{3}$ because

$(1.732)^2 = 2.999824$ □

We cannot say that 1.414 equals $\sqrt{2}$ but we say that it is *approximately equal* to it and write $1.414 \doteq \sqrt{2}$. Similarly, $1.732 \doteq \sqrt{3}$.

Table I provides decimal approximations of the square roots and cube roots of integers from 1 to 100. We can use this to obtain decimal approximations for irrational roots.

Example 3

$\sqrt{5} \doteq 2.236$ □

Example 4

$\sqrt[3]{7} \doteq 1.913$ □

Example 5

$\sqrt[3]{9} \doteq 2.080$ □

The simplification techniques developed in Sections 2 and 3 enable us to use this table to approximate more complicated roots.

Example 6

$\sqrt{5100} = \sqrt{100 \cdot 51} = \sqrt{100}\sqrt{51} = 10\sqrt{51} \doteq 10(7.141) = 71.41$

Thus $\sqrt{5100} \doteq 71.41$. □

Example 7

$\sqrt[3]{24,000} = \sqrt[3]{1000 \cdot 24} = 10\sqrt[3]{24} \doteq 10(2.884) = 28.84$

Thus $\sqrt[3]{24,000} \doteq 28.84$. □

TABLE I

n	$\sqrt{n}$	$\sqrt[3]{n}$	n	$\sqrt{n}$	$\sqrt[3]{n}$	n	$\sqrt{n}$	$\sqrt[3]{n}$	n	$\sqrt{n}$	$\sqrt[3]{n}$	n	$\sqrt{n}$	$\sqrt[3]{n}$
1	1.000	1.000	21	4.583	2.759	41	6.403	3.448	61	7.810	3.936	81	9.000	4.327
2	1.414	1.260	22	4.690	2.802	42	6.481	3.476	62	7.874	3.956	82	9.055	4.344
3	1.732	1.442	23	4.796	2.844	43	6.557	3.503	63	7.937	3.979	83	9.110	4.362
4	2.000	1.587	24	4.899	2.884	44	6.633	3.530	64	8.000	4.000	84	9.165	4.380
5	2.236	1.710	25	5.000	2.924	45	6.708	3.557	65	8.062	4.021	85	9.220	4.397
6	2.449	1.817	26	5.099	2.962	46	6.782	3.583	66	8.124	4.041	86	9.274	4.414
7	2.646	1.913	27	5.196	3.000	47	6.856	3.609	67	8.185	4.062	87	9.327	4.431
8	2.828	2.000	28	5.291	3.037	48	6.928	3.634	68	8.246	4.082	88	9.381	4.448
9	3.000	2.080	29	5.385	3.072	49	7.000	3.659	69	8.307	4.102	89	9.434	4.465
10	3.162	2.154	30	5.477	3.107	50	7.071	3.684	70	8.367	4.121	90	9.487	4.481
11	3.317	2.224	31	5.568	3.141	51	7.141	3.708	71	8.426	4.141	91	9.539	4.498
12	3.464	2.289	32	5.657	3.175	52	7.211	3.732	72	8.485	4.160	92	9.592	4.514
13	3.606	2.351	33	5.745	3.208	53	7.280	3.756	73	8.544	4.179	93	9.644	4.531
14	3.742	2.410	34	5.831	3.240	54	7.348	3.780	74	8.602	4.198	94	9.695	4.547
15	3.873	2.466	35	5.916	3.271	55	7.416	3.803	75	8.660	4.217	95	9.747	4.563
16	4.000	2.520	36	6.000	3.302	56	7.483	3.826	76	8.718	4.236	96	9.799	4.579
17	4.123	2.571	37	6.083	3.332	57	7.550	3.848	77	8.775	4.254	97	9.849	4.595
18	4.242	2.621	38	6.164	3.362	58	7.616	3.871	78	8.832	4.273	98	9.899	4.610
19	4.359	2.668	39	6.245	3.391	59	7.681	3.893	79	8.888	4.291	99	9.950	4.626
20	4.472	2.714	40	6.325	3.420	60	7.746	3.915	80	8.944	4.309	100	10.000	4.642

Example 8

$$\sqrt{.72} = \sqrt{(.01)(72)} = \sqrt{.01}\sqrt{72} = .1\sqrt{72} \doteq (.1)(8.485) = .8485$$

Thus $\sqrt{.72} \doteq .8485.$ □

The examples demonstrate that we can use the tables to obtain decimal approximations for square and cube roots of numbers that can be expressed as the product of a power of 10 and an integer between 1 and 100.

Our techniques for simplifying fractions involving roots are also helpful in approximating such expressions by decimals.

Example 9

$$\sqrt{\frac{5}{3}} = \sqrt{\frac{15}{9}} = \frac{\sqrt{15}}{3} \doteq \frac{3.873}{3} = 1.291 \quad \square$$

Example 10

$$\frac{\sqrt{7}}{\sqrt{2}} = \frac{\sqrt{7}\sqrt{2}}{\sqrt{2}\sqrt{2}} = \frac{\sqrt{14}}{2} \doteq \frac{3.742}{2} = 1.871 \quad \square$$

Example 11

$$\frac{2}{\sqrt{5} - \sqrt{2}} = \frac{2(\sqrt{5} + \sqrt{2})}{(\sqrt{5} - \sqrt{2})(\sqrt{5} + \sqrt{2})} = \frac{2(\sqrt{5} + \sqrt{2})}{3} \doteq \frac{2(2.236 + 1.414)}{3}$$

$$= \frac{2(3.650)}{3} = \frac{7.300}{3} \doteq 2.433 \quad \square$$

Example 12

$$\frac{\sqrt{3} + \sqrt{5}}{\sqrt{7} - \sqrt{3}} = \frac{(\sqrt{3} + \sqrt{5})(\sqrt{7} + \sqrt{3})}{(\sqrt{7} - \sqrt{3})(\sqrt{7} + \sqrt{3})} = \frac{\sqrt{21} + 3 + \sqrt{35} + \sqrt{15}}{7 - 3}$$

$$\doteq \frac{4.583 + 3 + 5.916 + 3.873}{4} = \frac{17.372}{4} = 4.343$$

Thus

$$\frac{\sqrt{3} + \sqrt{5}}{\sqrt{7} - \sqrt{3}} \doteq 4.343 \quad \square$$

The reader will appreciate the usefulness of simplification techniques if he determines the approximation in Example 12 directly by substituting approximations for the roots and then completing the computation.

We can solve some literal equations and formulas using roots and radicals and then use the tables to get approximate solutions or values.

Example 13

The area of a square is given by $A = s^2$ where s denotes the side of the square. Since $s > 0$, we know that $s = \sqrt{A}$ and we have a formula for the side of a square in terms of its area. If the area of a square is 47 sq in., then $s = \sqrt{47} = 6.856$. □

Example 14

The volume of a sphere is given by the formula $V = \frac{4}{3}\pi r^3$. Multiplying both sides by 3 yields

$$3V = 4\pi r^3$$

Dividing both sides by 4π yields

$$\frac{3V}{4\pi} = r^3$$

Since a number has only one cube root, we have

$$r = \sqrt[3]{\frac{3V}{4\pi}}$$

If the volume of the sphere is 11 cu in. and we approximate π by 22/7, then

$$r \doteq \sqrt[3]{\frac{3(11)}{4(22/7)}} = \sqrt[3]{\frac{21}{8}} = \frac{1}{2}\sqrt[3]{21} \doteq 1.379 \quad \square$$

EXERCISES

1. Determine decimal approximations for the following.
   (a) $\sqrt{43}$ (b) $\sqrt{92}$ (c) $\sqrt[3]{57}$
   (d) $\sqrt{13}$ (e) $\sqrt[3]{39}$ (f) $\sqrt{77}$

2. Determine decimal approximations for the following.
   (a) $\sqrt{5100}$ (b) $\sqrt{.67}$ (c) $\sqrt{.7}$
   (d) $\sqrt[3]{.092}$ (e) $\sqrt{8300}$ (f) $\sqrt[3]{.085}$
   (g) $\sqrt{228}$ (h) $\sqrt{275}$ (i) $\sqrt[3]{135}$
   (j) $\sqrt[3]{152}$ (k) $\sqrt{363}$ (l) $\sqrt[3]{875}$

3. Determine decimal approximations for the following.
   (a) $\sqrt{4700}$ (b) $\sqrt{.45}$ (c) $\sqrt{.8}$
   (d) $\sqrt[3]{.014}$ (e) $\sqrt{9600}$ (f) $\sqrt[3]{.046}$
   (g) $\sqrt{432}$ (h) $\sqrt{175}$ (i) $\sqrt[3]{104}$
   (j) $\sqrt[3]{297}$ (k) $\sqrt{405}$ (l) $\sqrt[3]{375}$

4. Determine decimal approximations for the following.
   (a) $\sqrt{7}/\sqrt{2}$ (b) $\sqrt[3]{13/2}$ (c) $\sqrt{19/3}$
   (d) $\sqrt{43/2}$ (e) $\sqrt{32/7}$ (f) $\sqrt[3]{3/4}$
   (g) $\sqrt[3]{5/3}$ (h) $\sqrt{21/2}$ (i) $\sqrt{11}/\sqrt{7}$
   (j) $\sqrt{17}/\sqrt{5}$ (k) $\sqrt[3]{13/8}$ (l) $\sqrt[3]{19}/\sqrt[3]{9}$

5. Determine decimal approximations for the following.

(a) $\sqrt{5}/\sqrt{3}$ (b) $\sqrt[3]{17/3}$ (c) $\sqrt{13/7}$

(d) $\sqrt{37/2}$ (e) $\sqrt{16/11}$ (f) $\sqrt[3]{7/4}$

(g) $\sqrt[3]{7/3}$ (h) $\sqrt{23/2}$ (i) $\sqrt{13}/\sqrt{7}$

(j) $\sqrt{12}/\sqrt{13}$ (k) $\sqrt[3]{17/8}$ (l) $\sqrt[3]{23}/\sqrt[3]{9}$

6. Determine decimal approximations for the following.

(a) $\dfrac{\sqrt{2}}{\sqrt{7}-\sqrt{5}}$ (b) $\dfrac{\sqrt{11}+\sqrt{3}}{\sqrt{11}-\sqrt{3}}$

(c) $\dfrac{\sqrt{10}-2\sqrt{5}}{\sqrt{10}+\sqrt{5}}$ (d) $\dfrac{\sqrt{6}+3\sqrt{2}}{\sqrt{7}-\sqrt{5}}$

(e) $\dfrac{\sqrt{19}-3\sqrt{5}}{\sqrt{19}+2\sqrt{5}}$ (f) $\dfrac{\sqrt[3]{3}+\sqrt[3]{2}}{\sqrt[3]{3}-\sqrt[3]{2}}$

7. Determine decimal approximations for the following.

(a) $\dfrac{\sqrt{5}}{\sqrt{7}-\sqrt{3}}$ (b) $\dfrac{\sqrt{13}+\sqrt{5}}{\sqrt{13}-\sqrt{5}}$

(c) $\dfrac{\sqrt{10}+3\sqrt{5}}{\sqrt{10}+\sqrt{5}}$ (d) $\dfrac{\sqrt{6}+2\sqrt{2}}{\sqrt{7}+\sqrt{5}}$

(e) $\dfrac{\sqrt{19}+4\sqrt{5}}{\sqrt{19}-3\sqrt{5}}$ (f) $\dfrac{\sqrt[3]{5}+\sqrt[3]{2}}{\sqrt[3]{5}-\sqrt[3]{2}}$

8. If a square has area 15 sq in., what is the length of its side to the nearest 1/100 of an inch?

9. If a sphere has volume of 176 cu in., what is its radius to the nearest 1/100 of an inch? (You may approximate π by 22/7.)

10. The area of a circle is given by $A = \pi r^2$. If the area of a circle is 55 sq in., what is a good approximation of its radius?

11. What is a good approximation for the edge of a cube that has a volume of 136 cu cm?

5. Rational Exponents

In Section 1 we extended the definition of exponents from positive integers to all integers. Thus a^z now has meaning if a is any nonzero real number and z is any integer. We made the definitions so that the laws of exponents would hold for the extended concept; in fact, they were made with these laws in mind. Now we wish to extend the concept of exponents to include all rational numbers as exponents, at least for powers of positive numbers. As before, we will keep the basic laws in mind and will formulate the definitions so that the laws will hold for the extended concept.

How would we expect to define $9^{1/2}$? If our previous laws are to hold we must have

$$(9^{1/2})^2 = 9^{1/2 \cdot 2} = 9^1 = 9$$

Thus if the laws are to be extended, $9^{1/2}$ must be a number whose square is 9. Hence $9^{1/2}$ must be 3 or -3. So far, we have not decided which one it is.

Consider also $(-5)^{1/2}$. If the laws of exponents are to be extended, we would have

$$[(-5)^{1/2}]^2 = (-5)^{1/2 \cdot 2} = (-5)^1 = -5$$

and $(-5)^{1/2}$ would be a number whose square is -5. But this is impossible. In general, if $a < 0$, we cannot assign a meaning to $a^{1/2}$ in a way that extends the laws of exponents since to do so would require a number whose square is negative and this is impossible.

In general, if $a \geq 0$, then for $a^{1/2}$ to have meaning and for the second law of exponents to hold, we must have

$$(a^{1/2})^2 = a^{1/2 \cdot 2} = a^1 = a$$

and $a^{1/2}$ must be a number whose square is a.

If $a > 0$, we know that two such numbers exist, namely $\sqrt{a}$ and $-\sqrt{a}$. We choose $a^{1/2}$ to be positive and make the following definition:

If $a \geq 0$, then $a^{1/2} = \sqrt{a}$.

Example 1

$5^{1/2} = \sqrt{5}$ □

Example 2

$4^{1/2} = 2$ □

Example 3

$12^{1/2} = 2\sqrt{3}$ □

By a line of reasoning analogous to that above, we cannot assign meaning to $(-5)^{1/4}$, $(-5)^{1/6}$, or $(-5)^{1/30}$ since to do so and extend the laws of exponents would require numbers with the property that an even power of them is positive; that is, if they have meaning then

$$[(-5)^{1/4}]^4 = -5, \quad [(-5)^{1/6}]^6 = -5 \quad \text{and} \quad [(-5)^{1/30}]^{30} = -5$$

No such numbers can exist because even powers are never negative.

With this thought in mind, we make the following definition which extends the meaning of $a^{1/2}$.

If n is an even positive integer and $a \geq 0$, then

$$a^{1/n} = \sqrt[n]{a}$$

What is the situation if n is odd? Again, if we wish to extend the second law of exponents, we must have

$$(a^{1/n})^n = a^{(1/n) \cdot n} = a^1 = a$$

Thus $a^{1/n}$ must be a number whose nth power is a. Since only one such number exists, the only way that we can make the definition consistent with the laws of exponents is to let $a^{1/n}$ be $\sqrt[n]{a}$. Thus we make the following definition.

If a is a real number and n is an odd positive integer, then

$$a^{1/n} = \sqrt[n]{a}$$

Example 4

$$5^{1/3} = \sqrt[3]{5} \ \square$$

Example 5

$$(-8)^{1/3} = -2 \ \square$$

Example 6

$$54^{1/3} = 3\sqrt[3]{2} \ \square$$

We now have meaning for $a^{1/n}$ if:

(i) $a \geq 0$ and n is a positive integer.
(ii) $a < 0$ and n is an odd positive integer.

Remember that $a^{1/n}$ has no meaning if $a < 0$ and n is even.

At this point, if n is a positive integer, we have given meaning to $a^{1/n}$ whenever it is possible to do so and not contradict the second law of exponents or the fact that even powers are nonnegative. Our object was to give meaning to rational exponents and thus far we have not considered fractional exponents other than $1/n$ with n in N.

If the laws of exponents are to be extended, then

$$5^{2/3} = 5^{2 \cdot 1/3} = (5^2)^{1/3} = \sqrt[3]{5^2}$$

and

$$5^{2/3} = 5^{1/3 \cdot 2} = (5^{1/3})^2 = (\sqrt[3]{5})^2$$

Thus if $5^{2/3}$ is to have meaning and the laws of exponents are to be extended, we must have

$$5^{2/3} = (5^{1/3})^2 = (5^2)^{1/3}$$

If $a \geq 0$ and p and q are positive integers, we define $a^{p/q}$ by

$$a^{p/q} = (\sqrt[q]{a})^p$$

Notice that

$$
\begin{aligned}
[(\sqrt[q]{a})^p]^q &= (\sqrt[q]{a})^{pq} \\
&= [(\sqrt[q]{a})^q]^p \\
&= (a)^p \\
&= a^p
\end{aligned}
$$

Since all numbers in the discussion are nonnegative and only one qth root of a^p exists,

$$(\sqrt[q]{a})^p = \sqrt[q]{a^p}$$

Thus for $a \geq 0$, we could just as well have defined $a^{p/q}$ to be $\sqrt[q]{a^p}$.

Example 7

$$4^{3/2} = (\sqrt[2]{4})^3 = 2^3 = 8 \quad \square$$

Example 8

$$12^{2/3} = \sqrt[3]{12^2} = \sqrt[3]{144} = \sqrt[3]{8}\,\sqrt[3]{18} = 2\sqrt[3]{18} \quad \square$$

Example 9

$$(.001)^{4/3} = (\sqrt[3]{.001})^4 = (.1)^4 = .0001 \quad \square$$

If our definition is adequate and precise, the meaning of $a^{p/q}$ should not depend on p and q but just the quotient. Thus we would expect $a^{2/3} = a^{4/6}$, and so on. To see that there is no difficulty, suppose p, q, and n are positive integers and $a \geq 0$. Then

$$(\sqrt[q]{a^p})^{qn} = [(\sqrt[q]{a^p})^q]^n = [a^p]^n = a^{pn}$$

Thus $\sqrt[q]{a^p}$ is the qnth root of a^{pn} and $\sqrt[q]{a^p} = \sqrt[qn]{a^{pn}}$. But

$$\sqrt[qn]{a^{pn}} = (\sqrt[qn]{a})^{pn} = a^{pn/qn}$$

Thus $a^{p/q}$ has the same meaning as $a^{pn/qn}$ and no difficulty arises.

Example 10

$$64^{2/3} = (\sqrt[3]{64})^2 = 4^2 = 16 \quad \square$$

Example 11

$$64^{4/6} = (\sqrt[6]{64})^4 = 2^4 = 16 \quad \square$$

Example 12

$$64^{8/12} = \sqrt[12]{64^8} = \sqrt[12]{(2^6)^8} = \sqrt[12]{2^{48}} = 2^4 = 16 \quad \square$$

Notice that *thus far* we have defined $a^{p/q}$ only if $p > 0$, $q > 0$, and $a \geq 0$. If $a < 0$, some difficulties arise in that $\sqrt[q]{a}$ may not exist. If q is odd, then $a^{p/q} = (\sqrt[q]{a})^p = \sqrt[q]{a^p}$ and no trouble arises.

If q is even and $a < 0$, then $\sqrt[q]{a}$ does not exist and so we cannot extend the definition in all cases. The real difficulty arises in the case of q odd, n even, and $pn/qn = p/q$. Then $\sqrt[q]{a}$ exists whereas $\sqrt[qn]{a}$ does not. To avoid difficulty we restrict the case for negative numbers to the case where q is odd or $p/q = s/t$ for some s and t with t odd. Thus if q is an odd positive integer, p is a positive integer, and $a < 0$,

$$a^{p/q} = (\sqrt[q]{a})^p$$

If q is even and p is odd, $a^{p/q}$ has no meaning. If q and p are both even, reduce p/q to lower terms with an odd denominator, if possible, and apply the definition. If this is not possible, then $a^{p/q}$ does not exist.

By an argument analogous to the previous one, it can be shown that

$$\sqrt[q]{a^p} = (\sqrt[q]{a})^p$$

if $a < 0$ and q is odd. If q and p are both even, the first expression has meaning but the second does not. Thus

If $a^{p/q}$ has meaning, then

$$a^{p/q} = (\sqrt[q]{a})^p = \sqrt[q]{a^p}$$

Our definition of the meaning of negative rational exponents is a natural extension of the meaning of negative integral exponents.

If p and q are positive integers and $a \neq 0$, then $a^{-p/q} = 1/a^{p/q}$ providing $a^{p/q}$ exists.

With these definitions we now have meaning for $a^{p/q}$ if p and q are integers with $q \neq 0$ and $a > 0$. We also have meaning if $q > 0$, q odd, and $a < 0$. Our definitions were chosen to conform to the four laws of exponents and, with some work, we can show that the laws do, in fact, hold. We will not include the proofs but suggest that the student attempt to prove some of them. We list the laws below for convenient reference.

If $a \in R$, $b \in R$, $x \in Q$, and $y \in Q$, then

$$a^x a^y = a^{x+y}$$

$$(a^x)^y = a^{xy}$$

$$(ab)^x = a^x b^x$$

$$\left(\frac{a}{b}\right)^x = \frac{a^x}{b^x}$$

providing all expressions given have meaning.

If $a > 0$ and $x \in R$, Axiom 12 may be used to define a meaning for a^x. If we let

$$M = \left\{ a^{p/q} : \frac{p}{q} \in Q \text{ and } \frac{p}{q} \le x \right\}$$

then a^x is defined to be the least upper bound of M, that is, the smallest number greater than or equal to every number in M. A great deal of work is involved in proving that such numbers exist and that the laws of exponents are extended by this definition. It can be done but will not be done in this text.

EXERCISES

1. Compute.
   (a) $4^{1/2}$ (b) $4^{-1/2}$
   (c) $4^{3/2}$ (d) $4^{-3/2}$
   (e) $(4^{2/3})^{3/2}$ (f) $(4^{-2/3})^{3/2}$
   (g) $4^{5/2}$ (h) $4^{-3/4}$

2. Compute.
   (a) $(-8)^{2/3}$ (b) $(-8)^{-2/3}$
   (c) $(-8)^{1/2}$ (d) $(-8)^{4/2}$
   (e) $(-8)^{4/6}$ (f) $(-8)^{6/4}$

3. Simplify.
   (a) $4^{2/3} 4^{-1/6}$ (b) $8^{3/4} 8^{-1/2}$
   (c) $(.01)^{1/6}(.01)^{1/3}$ (d) $\left(\frac{9}{4}\right)^{1/2}$
   (e) $\left(\frac{5}{8}\right)^{1/3}$ (f) $\left(\frac{8}{6}\right)^{6/4}$

4. Simplify.
   (a) $\left(\frac{4}{5}\right)^{1/2}\left(\frac{4}{5}\right)^{2/3}$ (b) $(.5)^{.9}(.5)^{-1.2}$
   (c) $\left(\frac{4}{3}\right)^{1/2}\left(\frac{25}{3}\right)^{1/2}$ (d) $\left(\frac{4}{3}\right)^{1/2}\left(\frac{4}{3}\right)^{1/3}$
   (e) $\left(\frac{1}{8}\right)^{1/2}\left(\frac{1}{8}\right)^{-1/6}$ (f) $\left(\frac{2}{9}\right)^{5/6} \div \left(\frac{2}{9}\right)^{1/3}$

5. Simplify.
   (a) $(4x^2 y^4)^{1/2}$ (b) $(20x^3 y^5)^{1/2}$
   (c) $(18x^7 y^8)^{1/2}$ (d) $(27x^7 y^8)^{1/3}$
   (e) $(4x^2 y^4)^{3/2}$ (f) $(8x^6 y^9)^{2/3}$
   (g) $(.008x^2)^{2/3}$ (h) $(400x^4)^{5/2}$
   (i) $(8x^{2/3} y^{4/5})^{4/3}$ (j) $(25x^{4/3} y^{5/2})^{1/2}$

6. Compute and simplify.

(a) $(2x^2y^5)^{1/3} \cdot (3x^4y^3)^{1/2}$

(b) $(2x^4y^5)^{1/3} \div (3x^2y^4)^{1/3}$

(c) $(3xy^7)^{1/3} \cdot (5x^2y^5)^{2/3}$

(d) $(4x^2y^3)^{2/5} \div (8xy^4)^{3/5}$

(e) $(16x^4y^3)^{3/5} \cdot (4x^8y^7)^{2/5}$

(f) $(8x^{1/3}y^{2/5})^2 \div (6x^{1/5}y^{2/3})^3$

(g) $(x^{2/3}y^{3/4})^6 \div (x^{3/4}y^{2/3})^5$

(h) $(8x^{5/4}y^{7/2})^{2/3} \div (4x^{3/2}y^{5/3})^{5/2}$

6. Summary of the Definitions and Laws Developed in this Chapter

1. $a^0 = 1$ if a is any number.

2. If a is a nonzero number and z is an integer, then

$$a^{-z} = \frac{1}{a^z}$$

3. If n is an even positive integer and $r \geq 0$, then $\sqrt[n]{r}$ is the nonnegative number whose nth power is r.

4. If n is an odd positive integer and r is a real number, then $\sqrt[n]{r}$ is the number whose nth power is r.

5. If q is a positive integer and r is a real number, then

$$r^{1/q} = \sqrt[q]{r}$$

providing $\sqrt[q]{r}$ exists.

6. If t is a rational number, r is a real number, and $t = p/q$, expressed in lowest terms, then

$$r^t = [\sqrt[q]{r}]^p$$

providing $\sqrt[q]{r}$ exists.

7. Let n be a positive integer and let a and b be real numbers such that $\sqrt[n]{a}$ and $\sqrt[n]{b}$ both exist. Then

$$(\sqrt[n]{a})(\sqrt[n]{b}) = \sqrt[n]{ab}$$

and

$$\sqrt[n]{\frac{a}{b}} = \frac{\sqrt[n]{a}}{\sqrt[n]{b}} \qquad (\text{providing } b \neq 0)$$

8. Let a and b be real numbers and x and y rational numbers. Then the following laws all hold (providing all expressions have meaning):

I. $a^x y^y = a^{x+y}$

II. $(a^x)^y = a^{xy}$

III. $(ab)^x = a^x b^x$

IV. $\left(\dfrac{a}{b}\right)^x = \dfrac{a^x}{b^x}$

9. A positive number is said to be expressed in scientific notation if it is expressed as the product of an integral power of 10 and a number between 1 and 10.

5

Polynomial Equations and Inequalities

In Chapter 2, we studied linear equations, linear inequalities, and systems of linear equations. Many problems, however, give rise to equations involving x^2, x^3, and so on. In this chapter we will be concerned with equations and inequalities involving polynomials in a single variable.

Since the equations are more complicated, the solution methods will be also. We will not present methods to solve *all* such equations and inequalities; indeed *no general method exists*. We will, however, develop techniques which enable us to solve many equations and, in particular, will develop a method of finding all solutions of quadratic equations. We will see many applications of these techniques to practical problems.

1. Solving Polynomial Equations by Factoring

An equation of the form

$$a_n x^n + a_{n-1} x^{n-1} + \cdots + a_1 x + a_0 = 0$$

with n a positive integer and each a_i a real number, is called a *polynomial* equation in one variable. Furthermore, any equation that can be transformed into such an equation by "adding the same thing to both sides" is also a polynomial equation in one variable.

153

Example 1

$5x + 3 = 0$ □

Example 2

$7x - 2 = 16 - 9$ □

Example 3

$4x^2 + 3x - 9 = 0$ □

Example 4

$(-3)x^4 + 6x^3 - 7x^2 + 3x - 18 = 0$ □

Example 5

$6x^2 + 2x = 4x^3 - 3x + 9$ □

In this section we will develop a method for solving polynomial equations using the techniques of factoring developed in Chapter 3. The method is based on the following rather simple principle which is, in fact, equivalent to Axiom 7.

If the product of two numbers is zero, then one of the numbers is zero.

Symbolically:

If $AB = 0$, then $A = 0$ or $B = 0$.

This principle also holds for the product of more than two numbers; that is, if the product of several numbers is zero, then at least one of the numbers must be zero.

Example 6

Solve the equation

$(x + 3)(x - 4) = 0$

If $(x + 3)(x - 4) = 0$ then one of the factors must be zero. Thus

$x + 3 = 0$ or $x - 4 = 0$

Hence

$x = -3$ or $x = 4$

Checking we see that

$$[(-3) + 3][(-3) - 4] = 0 \cdot (-7) = 0$$

and

$$[4 + 3][4 - 4] = (7) \cdot 0 = 0$$

Thus -3 and 4 are both solutions. There are no other solutions because if $x \neq -3$ and $x \neq 4$, then $x + 3 \neq 0$ and $x - 4 \neq 0$ and the product $(x + 3)(x - 4) \neq 0$. $\square$

Example 7

Solve the equation

$$7(2x + 3)(x - 1)(x + 5) = 0$$

For this product to be zero one of the factors must be zero. Since $7 \neq 0$,

$$7(2x + 3)(x - 1)(x + 5) = 0$$

if and only if

$$2x + 3 = 0, \qquad x - 1 = 0, \quad \text{or} \quad x + 5 = 0$$

Thus

$$x = -\tfrac{3}{2}, \qquad x = 1, \quad \text{or} \quad x = -5$$

The solution set is $\{-\tfrac{3}{2}, 1, -5\}$. $\square$

Example 8

Solve the equation

$$x^2 - 7x + 10 = 0$$

If

$$x^2 - 7x + 10 = 0$$

then

$$(x - 5)(x - 2) = 0$$

Thus

$$x - 5 = 0 \quad \text{or} \quad x - 2 = 0$$

and

$$x = 5 \quad \text{or} \quad x = 2$$

The solution set is $\{2, 5\}$. $\square$

Example 9

Solve the equation

$$x^2 + 3x = 4x + 6$$

First we add $-(4x + 6)$ to both sides and obtain the equation

$$x^2 - x - 6 = 0$$

Now one side is a polynomial and the other is zero. Factoring yields

$$(x - 3)(x + 2) = 0$$

Thus

$$x - 3 = 0 \quad \text{or} \quad x + 2 = 0$$

and

$$x = 3 \quad \text{or} \quad x = -2$$

The solution set is $\{3, -2\}$. $\square$

In order to solve a polynomial equation using factoring and the principle

If $AB = 0$, then $A = 0$ or $B = 0$,

we must have an equation in the form where one side is zero. Any equation can be transformed, by suitable additions, into one of this form. If the resulting equation is of the form

$$a_n x^n + a_{n-1} x^{n-1} + \cdots + a_1 x + a_0 = 0$$

we can attempt to solve the equation by factoring the polynomial and applying the principle. The procedure is as follows:

I. By suitable additions to both sides, obtain an equation in which one side is a polynomial and the other side is zero.

II. Rewrite the equation with the polynomial in completely factored form.

III. Set each factor equal to zero and solve the resulting equations. Any solution to one of these is a solution to the original equation and these are the only solutions to the original equation.

Remember that the method works only if one side .of the equation is zero.

Example 10

In the equation

$$(x - 7)(x + 5) = 1$$

there is no reason to expect

$$x - 7 = 1 \quad \text{or} \quad x + 5 = 1 \;\square$$

Study the following examples carefully before attempting the exercises. Make sure that you can justify each step.

Example 11

$$x^2 - 5x + 5 = 1$$
$$x^2 - 5x + 4 = 0$$
$$(x - 1)(x - 4) = 0$$
$$x - 1 = 0 \quad \text{or} \quad x - 4 = 0$$
$$x = 1 \quad \text{or} \quad x = 4 \;\square$$

Example 12

$$x^2 + 3x - 7 = 2x - 1$$
$$x^2 + x - 6 = 0$$
$$(x + 3)(x - 2) = 0$$
$$x + 3 = 0 \quad \text{or} \quad x - 2 = 0$$
$$x = -3 \quad \text{or} \quad x = 2 \;\square$$

Example 13

$$6x^2 - 19x + 10 = 0$$
$$(2x - 5)(3x - 2) = 0$$
$$2x - 5 = 0 \quad \text{or} \quad 3x - 2 = 0$$
$$2x = 5 \quad \text{or} \quad 3x = 2$$
$$x = \tfrac{5}{2} \quad \text{or} \quad x = \tfrac{2}{3} \;\square$$

Example 14

$$x^4 - 12x^2 = x^2 - 36$$
$$x^4 - 13x^2 + 36 = 0$$
$$(x^2 - 4)(x^2 - 9) = 0$$
$$(x + 2)(x - 2)(x + 3)(x - 3) = 0$$
$$x + 2 = 0, \quad x - 2 = 0, \quad x + 3 = 0, \quad \text{or} \quad x - 3 = 0$$

Thus

$$x = -2, \quad x = 2, \quad x = -3, \quad \text{or} \quad x = 3 \;\square$$

Example 15

$$40x^3 = 36x^2 - 8x$$

$$40x^3 - 36x^2 + 8x = 0$$

$$4x(10x^2 - 9x + 2) = 0$$

$$4x(2x - 1)(5x - 2) = 0$$

Since $4 \neq 0$, we must have

$$x = 0, \quad 2x - 1 = 0, \quad \text{or} \quad 5x - 2 = 0$$

Thus

$$x = 0, \quad x = \tfrac{1}{2}, \quad \text{or} \quad x = \tfrac{2}{5} \;\square$$

EXERCISES

1. Solve the following equations.
   (a) $x^2 - 5x - 6 = 0$ (b) $x^2 - 13x + 36 = 0$
   (c) $x^2 - 12x + 35 = 0$ (d) $x^2 - 6x - 7 = 0$
   (e) $x^2 + 2x - 15 = 0$ (f) $2x^2 - x - 3 = 0$
   (g) $6x^2 + 19x - 7 = 0$ (h) $x^2 - 4 = 0$

2. Solve the following equations.
   (a) $8x^2 + 10x = 3$ (b) $10x^2 = 33x + 7$
   (c) $x^2 - 6x = -9$ (d) $6x^2 - 97x + 18 = 2$
   (e) $42x^2 = 23x + 5$ (f) $2x^4 - 17x^3 + 21x^2 = 0$
   (g) $6x^3 - 5x^2 = x$ (h) $x^2 - 16 = 0$

3. Solve the following equations.
   (a) $2x^2 + 5x = 12$ (b) $18x^2 + 16x - 3 = 0$
   (c) $2x^3 - 13x^2 = 7x$ (d) $x^2 - 4x = 21$
   (e) $2x^3 = x^2 + 15x$ (f) $12x^3 = 8x^2 + 15x$
   (g) $x^2 - 4 = 0$ (h) $x^4 - 81 = 0$

4. Solve the following equations.
   (a) $2x^2 - 17x = 9$ (b) $2x^2 - 15 = x$
   (c) $x^3 + 21x = 10x^2$ (d) $x^2 = 10x + 132$
   (e) $6x^2 - 25x - 44 = 0$ (f) $30x^3 = 3x - 13x^2$
   (g) $x^2 - 9 = 0$ (h) $x^4 - 16 = 0$

5. Solve the following equations.
   (a) $x^2 = 25$ (b) $x^2 = 49$
   (c) $x^2 = \frac{1}{9}$ (d) $x^2 = 7$
   (e) $x^2 = \frac{7}{9}$ (f) $x^2 = k$ for some number $k \geq 0$.
   (g) $(x + 1)^2 = 25$ (h) $(x - 2)^2 = 49$
   (i) $(x + 3)^2 = \frac{1}{9}$ (j) $(x - 4)^2 = 7$
   (k) $(x + 5)^2 = \frac{7}{9}$ (l) $(x - p)^2 = k$ for some number $k \geq 0$ and some number p.

6. Solve the following equations.
   (a) $x^2 = 64$ (b) $x^2 = 81$
   (c) $x^2 = \frac{1}{4}$ (d) $x^2 = 11$
   (e) $x^2 = \frac{11}{4}$ (f) $x^2 = m$ for some number $m \geq 0$.

(g) $(x - 1)^2 = 64$ (h) $(x + 2)^2 = 81$
(i) $(x - 3)^2 = \frac{1}{4}$ (j) $(x + 4)^2 = 11$
(k) $(x - 5)^2 = \frac{11}{4}$ (l) $(x + q)^2 = m$ for some number $m \geq 0$ and some number q.

7. Solve the following equations.
   (a) $x^2 + 4x + 4 = 25$ (b) $x^2 - 6x + 9 = 16$
   (c) $x^2 + 6x + 9 = 5$ (d) $x^2 + 2xh + h^2 = a^2$
   (e) $x^2 - 2kx + k^2 = b^2$ (f) $x^2 + 2rx + r^2 = c$ with $c \geq 0$.

8. Solve the following equations.
   (a) $x^2 - 4x + 4 = 9$ (b) $x^2 + 6x + 9 = 25$
   (c) $x^2 + 4x + 4 = 7$ (d) $x^2 + 2ax + b^2 = k^2$
   (e) $x^2 - 2cx + c^2 = m^2$ (f) $x^2 - 2dx + d^2 = r$ with $r \geq 0$.

*9. Solve the following equations.
   (a) $x^4 - 13x^2 + 36 = 0$ (b) $x^4 - 17x^2 + 16 = 0$
   (c) $x^4 - 29x^2 + 100 = 0$ (d) $x^4 - (a^2 + b^2)x^2 + a^2b^2 = 0$

*10. Solve the following equations.
   (a) $x^4 - 5x^2 + 4 = 0$ (b) $x^4 - 25x^2 + 144 = 0$
   (c) $x^4 - 26x^2 + 25 = 0$ (d) $x^4 - (p^2 + q^2)x^2 + p^2q^2 = 0$

2. Completing the Square

A polynomial equation of the form

$$Ax^2 + Bx + C = 0 \qquad \text{with } A \neq 0$$

is called a *quadratic* equation. In the previous section we saw that many such equations can be solved using factoring. In the exercises you have probably experienced some of the frustration that can arise when trying to use the factorization method. In many cases no simple factorization exists and the method does not provide solutions. In this section we develop a technique that can be used to solve any quadratic equation, providing solutions exist. It also tells us when solutions do not exist. The method is called *completing the square*.

First recall from our study of square roots that a and b are numbers such that $a^2 = b^2$ if and only if $a = b$ or $a = -b$. If we have an equation in which both sides are perfect squares, then it is equivalent to two other equations.

Example 1

$$x^2 = 25$$

is equivalent to

$$x^2 = 5^2$$

which is equivalent to the pair of equations

$x = 5$ or $x = -5$

Thus the only solutions of $x^2 = 25$ are 5 and -5. $\square$

Example 2

The equation

$(x + 2)^2 = 9$

is equivalent to

$(x + 2)^2 = 3^2$

which is equivalent to the pair of equations

$x + 2 = 3$ or $x + 2 = -3$

Thus

$x = 1$ or $x = -5$ $\square$

Example 3

$(x - 4)^2 = 49$
$(x - 4)^2 = 7^2$
$x - 4 = 7$ or $x - 4 = -7$
$\qquad x = 11$ or $\qquad x = -3$ $\square$

Example 4

$(x - 3)^2 = 7$
$(x - 3)^2 = (\sqrt{7})^2$
$x - 3 = \sqrt{7}$ or $x - 3 = -\sqrt{7}$
$\qquad x = 3 + \sqrt{7}$ or $\qquad x = 3 - \sqrt{7}$ $\square$

Example 5

$(x + 5)^2 = 11$
$\qquad x + 5 = \sqrt{11}$ or $x + 5 = -\sqrt{11}$
$\qquad\qquad x = -5 + \sqrt{11}$ or $\qquad x = -5 - \sqrt{11}$ $\square$

From the examples, we see that for a general equation of the form

$(x + p)^2 = q^2$

we have

$x + p = q$ or $x + p = -q$

and

$$x = -p + q \quad \text{or} \quad x = -p - q$$

In general, if $r \geq 0$, then the equation

$$(x + p)^2 = r$$

is equivalent to the pair of equations

$$x + p = \sqrt{r} \quad \text{or} \quad x + p = -\sqrt{r}$$

and

$$x = -p + \sqrt{r} \quad \text{or} \quad x = -p - \sqrt{r}$$

The equation

$$(x + 3)^2 = -5$$

has no solutions because for all possible values for x,

$$(x + 3)^2 \geq 0$$

In general, if $r < 0$, the equation

$$(x + p)^2 = r$$

has no solution because $(x + p)^2 \geq 0$ for all possible values of x.

From the examples and the general form we see that if $r > 0$, equations of the form

$$(x + p)^2 = r$$

have two solutions. If $r = 0$, there is only one solution, $x = -p$. We now need to develop a method of transforming all solvable quadratic equations into an equation of this form.

Example 6

Solve the equation

$$x^2 + 2x = 8$$

The left side is not of the form $(x + p)^2$ or $(x - p)^2$ but if we add 1 to both sides we have

$$x^2 + 2x + 1 = 9$$

Then

$$(x + 1)^2 = 3^2$$
$$x + 1 = 3 \quad \text{or} \quad x + 1 = -3$$
$$x = 2 \quad \text{or} \quad x = -4$$

This equation could also have been solved by factoring

$$x^2 + 2x - 8 = 0$$
$$(x - 2)(x + 4) = 0$$
$$x - 2 = 0 \quad \text{or} \quad x + 4 = 0$$
$$x = 2 \quad \text{or} \quad x = -4 \; \square$$

Example 7

Solve the equation

$$x^2 + 4x = 7$$

The left side is not of the form $(x + p)^2$ or $(x - p)^2$ but if we add 4 to both sides of the equation we have

$$x^2 + 4x + 4 = 7 + 4$$

Then

$$(x + 2)^2 = 11$$

This is equivalent to

$$x + 2 = \sqrt{11} \quad \text{or} \quad x + 2 = -\sqrt{11}$$

Thus

$$x = -2 + \sqrt{11} \quad \text{or} \quad x = -2 - \sqrt{11}$$

Our knowledge of factoring would not have helped us to solve this equation. $\square$

Example 8

Solve the equation

$$x^2 - 6x = 5$$

The left side is not a perfect square but if we add 9 to both sides we have

$$x^2 - 6x + 9 = 14$$
$$(x - 3)^2 = 14$$

Thus

$$x - 3 = \sqrt{14} \quad \text{or} \quad x - 3 = -\sqrt{14}$$

and

$$x = 3 + \sqrt{14} \quad \text{or} \quad x = 3 - \sqrt{14} \; \square$$

In both of these cases we chose just the right number to add in order to make the left side into a perfect square. How did we decide what to add?

If we square a binomial of the form $(x + p)$ we get

$$(x + p)^2 = x^2 + 2px + p^2$$

Similarly,

$$(x - p)^2 = x^2 - 2px + p^2$$

In both cases, the constant p^2 is the square of $\frac{1}{2}$ of the coefficient of x. Thus if we have an equation of the form

$$x^2 + kx = m$$

we need to add $(k/2)^2$ to both sides.

$$x^2 + kx + \left(\frac{k}{2}\right)^2 = m + \left(\frac{k}{2}\right)^2$$

$$\left(x + \frac{k}{2}\right)^2 = m + \frac{k^2}{4}$$

Example 9

To transform the left side of

$$x^2 + 8x = 7$$

into a perfect square we add $\left(\frac{8}{2}\right)^2$ to both sides, that is, add 16 to both sides.

$$x^2 + 8x + 16 = 7 + 16$$
$$(x + 4)^2 = 23$$

Now we can proceed as before:

$$x + 4 = \sqrt{23} \quad \text{or} \quad x + 4 = -\sqrt{23}$$
$$x = -4 + \sqrt{23} \quad \text{or} \quad x = -4 - \sqrt{23} \quad \square$$

Example 10

$$x^2 - 7x = 5$$

We need to add $\left(-\frac{7}{2}\right)^2$ to both sides.

$$x^2 - 7x + \frac{49}{4} = 5 + \frac{49}{4}$$

$$\left(x - \frac{7}{2}\right)^2 = \frac{69}{4}$$

$$x - \frac{7}{2} = \sqrt{\frac{69}{4}} \quad \text{or} \quad x - \frac{7}{2} = -\sqrt{\frac{69}{4}}$$

$$x = \frac{7}{2} + \sqrt{\frac{69}{4}} \quad \text{or} \quad x = \frac{7}{2} - \sqrt{\frac{69}{4}}$$

$$x = \frac{7}{2} + \frac{\sqrt{69}}{2} \quad \text{or} \quad x = \frac{7}{2} - \frac{\sqrt{69}}{2}$$

$$x = \frac{7 + \sqrt{69}}{2} \quad \text{or} \quad x = \frac{7 - \sqrt{69}}{2} \quad \square$$

Example 11

$$x^2 + 11x = 4$$

$$x^2 + 11x + \left(\frac{11}{2}\right)^2 = 4 + \left(\frac{11}{2}\right)^2$$

$$\left(x + \frac{11}{2}\right)^2 = 4 + \frac{121}{4}$$

$$\left(x + \frac{11}{2}\right)^2 = \frac{137}{4}$$

$$x + \frac{11}{2} = \sqrt{\frac{137}{4}} \quad \text{or} \quad x + \frac{11}{2} = -\sqrt{\frac{137}{4}}$$

$$x = \frac{-11}{2} + \sqrt{\frac{137}{4}} \quad \text{or} \quad x = \frac{-11}{2} - \sqrt{\frac{137}{4}}$$

$$x = \frac{-11 + \sqrt{137}}{2} \quad \text{or} \quad x = \frac{-11 - \sqrt{137}}{2} \quad \square$$

Example 12

$$x^2 + 3x - 5 = 0$$

$$x^2 + 3x = 5$$

$$x^2 + 3x + \left(\frac{3}{2}\right)^2 = 5 + \left(\frac{3}{2}\right)^2$$

$$\left(x + \frac{3}{2}\right)^2 = 5 + \frac{9}{4}$$

$$\left(x + \frac{3}{2}\right)^2 = \frac{29}{4}$$

Adding 5 to both sides isolates terms with x on the left side.

$$x + \frac{3}{2} = \sqrt{\frac{29}{4}} \qquad \text{or} \qquad x + \frac{3}{2} = -\sqrt{\frac{29}{4}}$$

$$x = \frac{-3}{2} + \frac{\sqrt{29}}{2} \qquad \text{or} \qquad x = \frac{-3}{2} - \frac{\sqrt{29}}{2}$$

$$x = \frac{-3 + \sqrt{29}}{2} \qquad \text{or} \qquad x = \frac{-3 - \sqrt{29}}{2} \quad \square$$

Example 13

Solve

$$2x^2 - 4x = 13$$

In all our previous examples, the coefficient of x^2 was 1. If we divide both sides by 2, we obtain an equation in which the coefficient of x^2 is 1 and we proceed as before.

$$x^2 - 2x = \frac{13}{2}$$

$$x^2 - 2x + 1 = \frac{13}{2} + 1$$

$$(x - 1)^2 = \frac{15}{2}$$

$$x - 1 = \sqrt{\frac{15}{2}} \qquad \text{or} \qquad x - 1 = -\sqrt{\frac{15}{2}}$$

$$x = 1 + \sqrt{\frac{15}{2}} \qquad \text{or} \qquad x = 1 - \sqrt{\frac{15}{2}} \quad \square$$

The *completing the square technique* is a method by which we solve a quadratic equation by transforming it into an equivalent equation of the form

$$(x + p)^2 = r$$

The procedure is as follows:

A. By making suitable additions to both sides of the original equation, obtain an equation with terms involving only x^2 or x on one side and a constant as the other side.

B. By dividing or multiplying both sides by a suitable number, obtain an equation in which the coefficient of x^2 is 1.

C. Add the square of $\frac{1}{2}$ of the coefficient of x to both sides of the equation.

D. The resulting equation is of the form $(x + p)^2 = k$ which is equivalent to $x + p = \sqrt{k}$ or $x + p = -\sqrt{k}$ provided $\sqrt{k}$ exists. Solve these equations for x.

Example 14

$$4x^2 - 12x - 7 = 0$$
$$4x^2 - 12x = 7 \qquad \text{Adding 7 to both sides.}$$
$$x^2 - 3x = \tfrac{7}{4} \qquad \text{Dividing both sides by 4.}$$
$$x^2 - 3x + (-\tfrac{3}{2})^2 = \tfrac{7}{4} + (-\tfrac{3}{2})^2 \qquad \text{Adding } (-\tfrac{3}{2})^2 \text{ to both sides.}$$
$$(x - \tfrac{3}{2})^2 = \tfrac{7}{4} + \tfrac{9}{4}$$
$$(x - \tfrac{3}{2})^2 = \tfrac{16}{4}$$
$$(x - \tfrac{3}{2})^2 = 4$$

$$x - \tfrac{3}{2} = 2 \qquad \text{or} \qquad x - \tfrac{3}{2} = -2$$
$$x = \tfrac{3}{2} + 2 \quad \text{or} \qquad x = \tfrac{3}{2} - 2$$
$$x = \tfrac{7}{2} \qquad \text{or} \qquad x = -\tfrac{1}{2} \ \square$$

Example 15

$$2x^2 - 3x - 2 = 0$$
$$2x^2 - 3x = 2$$
$$x^2 - \tfrac{3}{2}x = 1$$
$$x^2 - \tfrac{3}{2}x + (-\tfrac{3}{4})^2 = 1 + (-\tfrac{3}{4})^2$$
$$(x - \tfrac{3}{4})^2 = 1 + \tfrac{9}{16}$$
$$(x - \tfrac{3}{4})^2 = \tfrac{25}{16}$$

$$x - \tfrac{3}{4} = \sqrt{\tfrac{25}{16}} \qquad \text{or} \qquad x - \tfrac{3}{4} = -\sqrt{\tfrac{25}{16}}$$
$$x - \tfrac{3}{4} = \tfrac{5}{4} \qquad \text{or} \qquad x - \tfrac{3}{4} = -\tfrac{5}{4}$$
$$x = \tfrac{3}{4} + \tfrac{5}{4} \quad \text{or} \qquad x = \tfrac{3}{4} - \tfrac{5}{4}$$
$$x = 2 \qquad \text{or} \qquad x = -\tfrac{1}{2} \ \square$$

Notice that Examples 14 and 15 could be solved by factoring. This is not surprising in that there is usually more than one way to solve a problem. The completing the square technique, however, is a somewhat automatic way of finding solutions whereas the factoring method involves "trial and error."

Example 16

$$3x^2 - 5x + 1 = 0$$
$$3x^2 - 5x = -1$$
$$x^2 - \frac{5}{3}x = -\frac{1}{3}$$
$$x^2 - \frac{5}{3}x + \left(-\frac{5}{6}\right)^2 = -\frac{1}{3} + \left(-\frac{5}{6}\right)^2$$
$$\left(x - \frac{5}{6}\right)^2 = -\frac{1}{3} + \frac{25}{36}$$

$$\left(x - \frac{5}{6}\right)^2 = -\frac{12}{36} + \frac{25}{36}$$

$$\left(x - \frac{5}{6}\right)^2 = \frac{13}{36}$$

$$x - \frac{5}{6} = \sqrt{\frac{13}{36}} \quad \text{or} \quad x - \frac{5}{6} = -\sqrt{\frac{13}{36}}$$

$$x = \frac{5}{6} + \frac{\sqrt{13}}{6} \quad \text{or} \quad x = \frac{5}{6} - \frac{\sqrt{13}}{6}$$

$$x = \frac{5 + \sqrt{13}}{6} \quad \text{or} \quad x = \frac{5 - \sqrt{13}}{6} \quad \square$$

Example 17

$$4x^2 - 5x - 8 = 0$$
$$4x^2 - 5x = 8$$
$$x^2 - \frac{5}{4}x = 2$$

$$x^2 - \frac{5}{4}x + \left(-\frac{5}{8}\right)^2 = 2 + \left(-\frac{5}{8}\right)^2$$

$$\left(x - \frac{5}{8}\right)^2 = 2 + \frac{25}{64}$$

$$\left(x - \frac{5}{8}\right)^2 = \frac{153}{64}$$

$$x - \frac{5}{8} = \frac{\sqrt{153}}{8} \quad \text{or} \quad x - \frac{5}{8} = -\frac{\sqrt{153}}{8}$$

$$x = \frac{5 + \sqrt{153}}{8} \quad \text{or} \quad x = \frac{5 + -\sqrt{153}}{8} \quad \square$$

Example 18

$$x^2 + x + 1 = 0$$
$$x^2 + x = -1$$
$$x^2 + x + (\tfrac{1}{2})^2 = -1 + (\tfrac{1}{2})^2$$
$$(x + \tfrac{1}{2})^2 = -1 + \tfrac{1}{4}$$
$$(x + \tfrac{1}{2})^2 = -\tfrac{3}{4}$$

This equation has no solution since $(x + \tfrac{1}{2})^2 \geq 0$ for all possible values of x, and therefore, cannot equal $-\tfrac{3}{4}$. $\square$

In general, if the "completing the square" technique produces an equation of the form

$$(x + p)^2 = r \qquad \text{with } r < 0$$

then the equation has no solutions because

$$(x + p)^2 \geq 0$$

for all values of x.

The technique of completing the square is not only used for solving quadratic equations, it also is of importance in graphing, as we will see in Chapter 6, and in other areas of mathematics.

EXERCISES

Solve the following equations (if they have solutions) by completing the square. Simplify and check your answers.

1. $x^2 + 4x = 8$
2. $x^2 - 6x = 4$
3. $x^2 + 2x = 19$
4. $x^2 - 10x = 11$
5. $x^2 - 3x = 2$
6. $x^2 + 5x = 3$
7. $x^2 + 8x - 2 = 0$
8. $x^2 - 5x + 3 = 0$
9. $2x^2 + 6x - 5 = 0$
10. $3x^2 + 2x - 1 = 0$
11. $5x^2 + 4x + 10 = 0$
12. $-3x^2 - 9x + 5 = 0$
13. $\frac{1}{4}x^2 + 2x - 1 = 0$
14. $7x^2 - 3x - 5 = 0$
15. $11x^2 - 13x + 2 = 0$
16. $5x^2 + 2x - 10 = 0$
17. $4x^2 + 3x - 1 = 0$
18. $3x^2 - 4 = 5x$
19. $12x^2 + 3x - 36 = 0$
20. $\frac{1}{3}x^2 + \frac{1}{6}x - \frac{1}{2} = 0$
21. $\frac{1}{4}x^2 - \frac{2}{3}x - 3 = 0$
22. $15x^2 - 4x - 3 = 0$
23. $2x^2 + 13x = 2$
24. $8x^2 + 5x - 2 = 0$
25. $3x^2 + 4x + 5 = 0$
26. $8x^2 - 5x + 2 = 0$
27. $6x^2 + 3x - 7 = 0$
28. $.01x^2 + 2x - 1 = 0$
29. $.3x^2 + 4x - .2 = 0$
30. $18x^2 + 45x - 5 = 0$

*31. Use the completing the square technique to solve the general equation

$$Ax^2 + Bx + C = 0$$

You may assume $A \neq 0$.

3. The Quadratic Formula

Once a general method has been devised for solving equations of a particular form, the method can be applied to obtain a formula that provides all such solutions. In this section we apply the technique of completing the square to the general quadratic equation

$$Ax^2 + Bx + C = 0 \qquad (A \neq 0)$$

to provide a general solution or formula for such equations. (You were asked to do so in Exercise 31 of Section 2.)

$$Ax^2 + Bx + C = 0$$
$$Ax^2 + Bx = -C$$
$$x^2 + \frac{B}{A}x = \frac{-C}{A}$$

We now add $(B/2A)^2$ to both sides to complete the square.

$$x^2 + \frac{B}{A}x + \left(\frac{B}{2A}\right)^2 = \left(\frac{B}{2A}\right)^2 - \frac{C}{A}$$

$$\left(x + \frac{B}{2A}\right)^2 = \frac{B^2}{4A^2} - \frac{C}{A}$$

$$\left(x + \frac{B}{2A}\right)^2 = \frac{B^2}{4A^2} - \frac{4AC}{4A^2}$$

$$\left(x + \frac{B}{2A}\right)^2 = \frac{B^2 - 4AC}{4A^2}$$

Thus

$$x + \frac{B}{2A} = \sqrt{\frac{B^2 - 4AC}{4A^2}} \quad \text{or} \quad x + \frac{B}{2A} = -\sqrt{\frac{B^2 - 4AC}{4A^2}}$$

Simplifying the right sides we have

$$x + \frac{B}{2A} = \frac{\sqrt{B^2 - 4AC}}{2A} \quad \text{or} \quad x + \frac{B}{2A} = \frac{-\sqrt{B^2 - 4AC}}{2A}$$

Then

$$x = \frac{-B}{2A} + \frac{\sqrt{B^2 - 4AC}}{2A} \quad \text{or} \quad x = \frac{-B}{2A} - \frac{\sqrt{B^2 - 4AC}}{2A}$$

$$x = \frac{-B + \sqrt{B^2 - 4AC}}{2A} \quad \text{or} \quad x = \frac{-B - \sqrt{B^2 - 4AC}}{2A}$$

These equations are usually condensed to

$$x = \frac{-B \pm \sqrt{B^2 - 4AC}}{2A}$$

where the $\pm$ sign indicates the two equations.

The above equation is known as the *quadratic formula* and can be used to find all solutions to quadratic equations if they exist and to indicate those that have no solutions. *The student should memorize the formula.*

Example 1

Solve

$$x^2 + 2x - 5 = 0$$

Using the quadratic formula yields

$$x = \frac{-2 \pm \sqrt{4 + 20}}{2} \qquad \text{Since } A = 1, B = 2, \text{ and } C = -5.$$

$$x = \frac{-2 \pm \sqrt{24}}{2}$$

$$x = \frac{-2 \pm 2\sqrt{6}}{2}$$

$$x = -1 \pm \sqrt{6}$$

$$x = -1 + \sqrt{6} \quad \text{or} \quad x = -1 - \sqrt{6} \ \square$$

Example 2

Solve

$$3x^2 - 12x + 7 = 0$$

Applying the formula yields

$$x = \frac{12 \pm \sqrt{144 - 84}}{6} \qquad \text{Since } A = 3, B = -12,$$
$$\text{and } C = 7.$$

$$x = \frac{12 \pm \sqrt{60}}{6}$$

$$x = \frac{12 \pm 2\sqrt{15}}{6}$$

$$x = 2 \pm \frac{\sqrt{15}}{3}$$

$$x = 2 + \frac{\sqrt{15}}{3} \quad \text{or} \quad x = 2 - \frac{\sqrt{15}}{3} \ \square$$

Example 3

If we try to apply the formula to

$$x^2 + 3x + 5 = 0$$

we obtain

$$x = \frac{-3 \pm \sqrt{9 - 20}}{2}$$

$$x = \frac{-3 \pm \sqrt{-11}}{2}$$

This is impossible because $\sqrt{-11}$ does not exist. □

The equation

$$Ax^2 + Bx + C = 0 \qquad \text{with } A \neq 0$$

has no solution if $B^2 - 4AC$ is negative because $\sqrt{B^2 - 4AC}$ does not exist. The number $B^2 - 4AC$ is called the *discriminant* of the quadratic equation. In summary:

The equation

$$Ax^2 + Bx + C = 0 \qquad \text{with } A \neq 0$$

has, as its only solutions

$$\frac{-B \pm \sqrt{B^2 - 4AC}}{2A}$$

The equation has *two* solutions if the discriminant $B^2 - 4AC$ is positive. The equation has *one* solution if the discriminant is 0. The equation has *no* solution if the discriminant is negative.

EXERCISES

1. Solve the following using the quadratic formula.
   (a) $x^2 + 4x - 8 = 0$ (b) $x^2 - 7x - 3 = 0$
   (c) $3x^2 + x - 2 = 0$ (d) $2x^2 + 12x = 5$
   (e) $2x^2 + 3x = 5$ (f) $3x^2 + 2x - 7 = 0$
   (g) $4x^2 + x - 2 = 0$ (h) $3x^2 + 2x = 1$
2. Solve the following using the quadratic formula.
   (a) $x^2 + 5x - 7 = 0$ (b) $x^2 + 9x + 2 = 0$
   (c) $4x^2 + 5x - 3 = 0$ (d) $3x^2 + 5x = 8$
   (e) $3x^2 + 7x - 4 = 0$ (f) $5x^2 + 3x = 9$
   (g) $4x^2 + 5x - 3 = 0$ (h) $6x^2 + 2x - 1 = 0$
3. Use the quadratic formula to solve the following equations (if they have solutions).
   (a) $2x^2 + 3x - 4 = 0$ (b) $x^2 + 5x - 2 = 0$
   (c) $7x^2 + x - 9 = 0$ (d) $2x^2 + \frac{1}{3}x - \frac{1}{4} = 0$
   (e) $5x^2 + 7x + 1 = 0$ (f) $\sqrt{7}x^2 + \sqrt{5}x - \sqrt{7} = 0$
   (g) $14x^2 + 17x - 32 = 0$ (h) $16x^2 - 15x + 3 = 0$
   (i) $11x^2 + 12x + 13 = 0$ (j) $1.3x - 2.4x + 1.2 = 0$

4. Use the quadratic formula to solve the following equations (if they have solutions).

(a) $x^2 + x + 1 = 0$

(b) $2x^2 + 3x - 5 = 0$

(c) $8x^2 - 13x + 2 = 0$

(d) $3x^2 + \frac{1}{4}x - \frac{1}{5} = 0$

(e) $17x^2 + 5x + 1 = 0$

(f) $\sqrt{3}x^2 + \sqrt{2}x - \sqrt{3} = 0$

(g) $18x^2 - 14x - 13 = 0$

(h) $23x^2 - 5x - 20 = 0$

(i) $12x^2 + 11x - 13 = 0$

(j) $1.6x^2 - 1.3x - 1.7 = 0$

5. By direct multiplication, verify that

$$Ax^2 + Bx + C = A \left[x + \frac{B - \sqrt{B^2 - 4AC}}{2A} \right] \left[x + \frac{B + \sqrt{B^2 - 4AC}}{2A} \right]$$

6. Explain why the equation $Ax^2 + Bx + C = 0$ will always have a solution if $A > 0$ and $C < 0$.

7. Use the assertion from Exercise 5 to factor the following expressions.

(a) $x^2 + 5x - 10$

(b) $3x^2 + 2x - 7$

(c) $5x^2 - 3x - 1$

(d) $16x^2 - 17x + 2$

(e) $7x^2 + 3x - 19$

(f) $4x^3 + 7x^2 + x$

(g) $5x^2 - 3x + 2$

(h) $.4x^2 + 5x - 1.2$

8. Use the assertion from Exercise 5 to establish the following expressions.

(a) $x^2 - 6x - 10$

(b) $2x^2 + 3x - 5$

(c) $4x^2 - 3x - 1$

(d) $18x^2 + 19x + 2$

(e) $9x^2 + 5x - 7$

(f) $6x^2 + 10x + 2$

(g) $5x^2 + 3x - 2$

(h) $.6x^2 + 8x = 1.3$

9. Determine the average of the solutions to each of the equations in Exercise 2; that is, if the solutions are 3 and 11, the average is 7.

10. Use the quadratic formula to determine the average of the solutions of the general quadratic formula $Ax^2 + Bx + C = 0$.

11. Use your result from Exercise 10 to determine the average of the solutions to each of the equations in Exercise 4.

4. Polynomial Inequalities

Just as we needed to extend our ability to solve equations from linear equations to polynomial equations, we need to extend our knowledge of linear inequalities to polynomial inequalities. The basic method to be developed uses factoring to reduce a polynomial inequality to a *system* of linear inequalities such as those studied in Chapter 2.

The basic method is based on the following rather simple principles that follow immediately from Theorem 7 in Chapter 1.

I. The product of positive numbers is positive.

II. The product of an even number of negative numbers is positive.

III. The product of an odd number of negative numbers is negative.

In summary, if $A_1, \ldots, A_n$ are numbers, then the product

$$A_1 \cdot A_2 \cdot \ldots \cdot A_n$$

is *positive* if and only if an *even* number of the factors are negative and is *negative* if and only if an *odd* number of the factors are negative.

Example 1

Solve the inequality

$$(x - 1)(x - 3) > 0$$

In order for $(x - 1)(x - 3)$ to be positive, either both factors are positive or both factors are negative.

Case 1

$$x - 1 > 0 \quad \text{and} \quad x - 3 > 0$$
$$x > 1 \quad \text{and} \quad x \quad\quad > 3$$

Thus $x > 3$.

Case 2

$$x - 1 < 0 \quad \text{and} \quad x - 3 < 0$$
$$x < 1 \quad \text{and} \quad x \quad\quad < 3$$

Thus $x < 1$.

The solution set is $\{x : x < 1 \text{ or } x > 3\}$ (Figure 5-1). ☐

Figure 5-1 Solution set of $(x - 1)(x - 3) > 0$

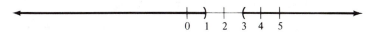

Example 2

Solve

$$(x - 5)(2x + 1)(3x - 2) > 0$$

This product is positive if all factors are positive or one is positive and two are negative.

Case 1

$$x - 5 > 0, \quad 2x + 1 > 0, \quad \text{and} \quad 3x - 2 > 0$$
$$x > 5, \quad\quad x > -\tfrac{1}{2}, \quad \text{and} \quad\quad x > \tfrac{2}{3}$$

Thus the solution set for Case 1 is $\{x : x > 5\}$.

Case 2

$$x - 5 < 0, \quad 2x + 1 < 0, \quad \text{and} \quad 3x - 2 > 0$$
$$x < 5, \quad\quad x < -\tfrac{1}{2}, \quad \text{and} \quad\quad x > \tfrac{2}{3}$$

Since $x < -\tfrac{1}{2}$ and $x > \tfrac{2}{3}$ is impossible, Case 2 produces no solutions.

Case 3

$$x - 5 < 0, \quad 2x + 1 > 0, \quad \text{and} \quad 3x - 2 < 0$$
$$x < 5, \quad\quad x > -\tfrac{1}{2}, \quad \text{and} \quad\quad x < \tfrac{2}{3}$$

Thus Case 3 yields $\{x : -\tfrac{1}{2} < x < \tfrac{2}{3}\}$ as the solution set.

Case 4

$$x - 5 > 0, \quad 2x + 1 < 0, \quad \text{and} \quad 3x - 2 < 0$$
$$x > 5, \quad\quad x < -\tfrac{1}{2}, \quad \text{and} \quad\quad x < \tfrac{2}{3}$$

Since $x > 5$ and $x < -\tfrac{1}{2}$ is impossible, Case 4 produces no solutions. Thus $\{x : -\tfrac{1}{2} < x < \tfrac{2}{3} \text{ or } 5 < x\}$ is the solution set (Figure 5-2).

The solution set is the union of all the solution sets arising in each case. □

Figure 5-2 Solution set of $(x - 5)(2x + 1)(3x - 2) > 0$

Example 3

Solve

$$x^2 - 2x - 3 < 0$$

Factoring yields

$$(x - 3)(x + 1) < 0$$

In order for this product to be negative, one factor must be positive and the other negative.

Case 1

$x - 3 < 0$ and $x + 1 > 0$
$x < 3$ and $x > -1$

that is, $-1 < x < 3$.

Case 2

$x - 3 > 0$ and $x + 1 < 0$
$x > 3$ and $x < -1$

This is impossible.
Thus the solution set is $\{x: -1 < x < 3\}$ (Figure 5-3). □

Figure 5-3 Solution set of $x^2 - 2x - 3 < 0$

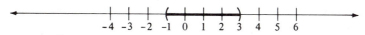

Example 4

Solve $x^2 - 5x + 6 > 0$. Factoring yields

$(x - 3)(x - 2) > 0$

Case 1

$x - 3 > 0$ and $x - 2 > 0$
$x > 3$ and $x > 2$

Hence $x > 3$.

Case 2

$x - 3 < 0$ and $x - 2 < 0$
$x < 3$ and $x < 2$

Hence $x < 2$.
The solution set is $\{x: x < 2 \text{ or } x > 3\}$ (Figure 5-4). □

Figure 5-4 Solution set of $x^2 - 5x + 6 > 0$

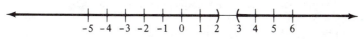

Example 5

Solve $x^2 - 7x > 8$.

$$x^2 - 7x - 8 > 0$$ 　　　　　　Subtracting 8 from both sides.
$$(x - 8)(x + 1) > 0$$ 　　　　　　Factoring.

Case 1

$x - 8 > 0$　and　$x + 1 > 0$
　　$x > 8$　and　　　$x > -1$

Hence $x > 8$.

Case 2

$x - 8 < 0$　and　$x + 1 < 0$
　　$x < 8$　and　　　$x < -1$

Hence $x < -1$.
　　The solution set is $\{x: x < -1 \text{ or } x > 8\}$ (Figure 5-5). □

Figure 5-5　Solution set of $x^2 - 7x > 8$

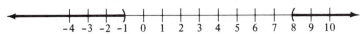

The general technique for solving polynomial inequalities is as follows:

    I. By suitable additions or subtractions transform the original inequality into one in which one side is zero and the other side is a polynomial.
    II. Factor the polynomial.
   III. Determine *all systems of inequalities* which are compatible with the original inequality; that is, *consider all possible cases.*
   IV. Solve the resulting systems of inequalities.
    V. The solution set is the union of the solution sets of each of the systems.

EXERCISES

1. Solve the following inequalities. Draw graphs of the solution sets.
    (a) $(x + 4)(x - 5) < 0$ 　　　　(b) $(x - 7)(x + 2) > 0$
    (c) $(2x + 3)(x - 4) > 0$ 　　　　(d) $x(x + 3)(x - 2) < 0$
    (e) $x^2(x - 1)(x + 4) > 0$ 　　　　(f) $x^2 - 7x + 10 > 0$

(g) $x^2 + 9x < 20$ (h) $x^2 + 6x > 7$
(i) $2x^2 - 9x < 35$ (j) $x^3 + 8x^2 - 20x > 0$
(k) $x^2 - 4x + 4 > 0$ (l) $x^2 + 6x + 9 < 0$

2. Solve the following inequalities. Draw graphs of the solution sets.
  (a) $(x + 6)(x - 5) > 0$ (b) $(x - 9)(x + 2) < 0$
  (c) $(3x - 1)(x + 5) > 0$ (d) $x(x + 7)(x - 11) < 0$
  (e) $x^2(x + 3)(x - 9) > 0$ (f) $x^2 - 10x - 24 < 0$
  (g) $x^2 + 7x > 18$ (h) $x^2 + 12x > -32$
  (i) $6x^2 + x < 15$ (j) $x^3 - 9x^2 + 18x < 0$
  (k) $x^2 + 10x + 25 > 0$ (l) $x^2 - 14x + 49 < 0$

3. Solve the following inequalities. Draw graphs of the solution sets.
  (a) $(x + 2)x(x - 3) > 0$
  (b) $(x + 1)(x - 1)(x - 2) < 0$
  (c) $(x + 7)(x + 5)(x - 4)(x - 9) > 0$
  (d) $(x + 6)(x + 3)(x - 7)(x - 11) < 0$
  (e) $(x + 9)(x + 6)(x + 2)x(x - 4)(x - 5) > 0$
  (f) $x^2(x + 5)(x - 7) > 0$
  (g) $(x + 2)^2(x + 3)(x - 4) > 0$
  (h) $(x - 1)^2(x + 7)(x - 8) < 0$
  (i) $(x + 5)^2(x - 7)^4(x + 2)(x - 3) > 0$
  (j) $(x + 2)^2(x - 3)^2(x - 7)^2 > 0$

4. Solve the following inequalities. Draw graphs of the solution sets.
  (a) $(x + 3)x(x - 2) < 0$
  (b) $(x + 2)(x - 2)(x - 5) > 0$
  (c) $(x + 5)(x + 2)(x - 3)(x - 7) > 0$
  (d) $(x + 8)(x + 1)(x - 7)(x - 11) < 0$
  (e) $(x + 10)(x + 5)(x + 3)(x)(x - 7)(x - 10) > 0$
  (f) $x^2(x + 6)(x - 9) < 0$
  (g) $(x - 1)^2(x + 5)(x - 2) > 0$
  (h) $(x + 4)^2(x + 10)(x - 9) < 0$
  (i) $(x + 7)^4(x - 2)^2(x + 8)(x - 5) > 0$
  (j) $(x + 5)^2(x - 4)^2(x - 11)^2 > 0$

5. (a) Solve the inequality $(x - p)(x - q) > 0$ if $p < q$.
  (b) Solve the inequality $(x - p)(x - q) < 0$ if $p < q$.
  (c) Draw graphs of the solution sets for parts (a) and (b).

6. (a) Solve the inequality $(x - p)(x - q)(x - r) > 0$ if $p < q < r$.
  (b) Solve the inequality $(x - p)(x - q)(x - r) < 0$ if $p < q < r$.
  (c) Draw graphs of the solution sets for parts (a) and (b).

7. (a) Solve the inequality $(x - p)^2(x - q)(x - r) > 0$ if $q < r$.
  (b) Solve the inequality $(x - p)^2(x - q)(x - r) < 0$ if $q < r$.

*8. Let $p_1, p_2, \ldots, p_n$ be numbers such that $p_1 < p_2 < \cdots < p_n$.
  (a) Solve the inequality

$$(x - p_1)(x - p_2)(x - p_3) \ldots (x - p_n) > 0, \qquad \text{if } n \text{ is even}$$

  (b) Solve the same inequality if n is odd.

5. Solving Quadratic Inequalities

The general techniques for solving quadratic equations can be modified for inequalities involving quadratic polynomials even if they do not factor readily.

Notice first that if a and b are numbers such that

$$a^2 > b^2$$

then

$$|a| > |b|$$

If, in fact, b is positive, we have

$$|a| > b$$

Thus, either $a > b$ or $-a > b$, and therefore $a > b$ or $a < -b$. Thus, if b is positive, the inequality

$$a^2 > b^2$$

is equivalent to the system

$$a > b \quad \text{or} \quad a < -b$$

Example 1

Using the fact just established, we see that the inequality

$$x^2 > 4$$

is equivalent to the system

$$x > 2 \quad \text{or} \quad x < -2 \quad \square$$

Example 2

The inequality

$$x^2 > 7$$

is equivalent to

$$x > \sqrt{7} \quad \text{or} \quad x < -\sqrt{7} \quad \square$$

Example 3

By the same reasoning, the inequality

$$(x + 3)^2 > 4$$

is equivalent to

$(x + 3) > 2$ or $(x + 3) < -2$

which means that

$x > -1$ or $x < -5$ ☐

Example 4

The inequality

$(x - 5)^2 > 9$

is equivalent to the system

$(x - 5) > 3$ or $(x - 5) < -3$
$x > 8$ or $\phantom{(x-5)<}x < 2$ ☐

Example 5

The inequality

$(2x + 1)^2 > 25$

is equivalent to the system

$(2x + 1) > 5$ or $(2x + 1) < -5$

Thus,

$2x > 4$ or $2x < -6$
$x > 2$ or $x < -3$ ☐

Example 6

The inequality

$(x + 4)^2 > 7$

is equivalent to the system

$(x + 4) > \sqrt{7}$ or $(x + 4) < -\sqrt{7}$
$x > \sqrt{7} - 4$ or $\phantom{(x+4)<}x < -\sqrt{7} - 4$

Again recall that if

$a^2 > b^2$

then

$|a| > |b|$

Thus if a is positive, we have

$a > |b|$

This inequality, we know, is equivalent to

$-a < b < a$

Thus if $a > 0$, the inequality

$b^2 < a^2$

is equivalent to the system

$-a < b < a$ ☐

Example 7

Using this fact, the inequality

$x^2 < 4$

is equivalent to

$-2 < x < 2$ ☐

Example 8

Using the same fact, the inequality

$x^2 < 7$

is equivalent to

$-\sqrt{7} < x < \sqrt{7}$ ☐

Example 9

The inequality

$(x + 2)^2 < 9$

is equivalent to

$-3 < x + 2 < 3$

that is, to the system

$-3 < x + 2$ and $x + 2 < 3$

Thus,

$-5 < x$ and $x < 1$ ☐

Example 10

The inequality

$$(x - 5)^2 < 3$$

is equivalent to

$$-\sqrt{3} < x - 5 < \sqrt{3}$$

which means that

$$5 - \sqrt{3} < x < 5 + \sqrt{3} \quad \square$$

Example 11

The inequality

$$(x + 7)^2 < 15$$

is equivalent to

$$-\sqrt{15} < x + 7 < \sqrt{15}$$

which means that

$$-7 - \sqrt{15} < x < -7 + \sqrt{15} \quad \square$$

Example 12

The inequality

$$(3x + 2)^2 < 5$$

is equivalent to

$$-\sqrt{5} < 3x + 2 < \sqrt{5}$$

Thus,

$$-\sqrt{5} - 2 < 3x < \sqrt{5} - 2$$

and

$$\frac{-\sqrt{5} - 2}{3} < x < \frac{\sqrt{5} - 2}{3} \quad \square$$

The principles and examples just given provide us with a way of solving quadratic inequalities once we have a quadratic inequality in which one side is the square of a binomial and the other side is a number. The completing the square technique can be used to transform quadratic inequalities into inequalities of this general form.

Example 13

Solve

$$x^2 + 4x - 7 < 0$$

Adding 7 to both sides yields

$$x^2 + 4x < 7$$

To complete the square on the left side we need to add 4 to both sides. Thus, we have

$$x^2 + 4x + 4 < 11$$
$$(x + 2)^2 < 11$$

Then

$$-\sqrt{11} < x + 2 < \sqrt{11}$$

and

$$-2 - \sqrt{11} < x < -2 + \sqrt{11} \quad \square$$

Example 14

Solve

$$x^2 - 6x > 5$$

Adding 9 to both sides yields

$$x^2 - 6x + 9 > 14$$

Thus,

$$(x + 3)^2 > 14$$

Hence,

$$x + 3 > \sqrt{14} \qquad \text{or} \quad x + 3 < -\sqrt{14}$$
$$x > -3 + \sqrt{14} \quad \text{or} \qquad x < -3 - \sqrt{14}$$

Example 15

Solve

$$2x^2 + 3x - 8 < 0$$

Adding 8 to both sides yields

$$2x^2 + 3x < 8$$

Dividing both sides by 2, we have

$$x^2 + \tfrac{3}{2}x < 4$$

To complete the square on the left side, we add $(\frac{3}{4})^2$ to both sides

$$x^2 + \tfrac{3}{2}x + (\tfrac{3}{4})^2 < 4 + \tfrac{9}{16}$$

Then

$$\left(x + \frac{3}{4}\right)^2 < \frac{57}{16}$$

Thus,

$$\frac{-\sqrt{57}}{4} < x + \frac{3}{4} < \frac{\sqrt{57}}{16}$$

$$\frac{-3}{4} - \frac{\sqrt{57}}{4} < \quad x \quad < \frac{-3}{4} + \frac{\sqrt{57}}{16}$$

or

$$\frac{-3 - \sqrt{57}}{4} < \quad x \quad < \frac{-3 + \sqrt{57}}{4} \quad \square$$

Example 16

Solve

$$3x^2 + 4x - 2 > 0$$

Adding 2 to both sides yields

$$3x^2 + 4x > 2$$

Dividing both sides by 3 yields

$$x^2 + \frac{4}{3}x > \frac{2}{3}$$

To complete the square on the left side, we add $(\frac{2}{3})^2$ to both sides.

$$x^2 + \frac{4}{3}x + \left(\frac{2}{3}\right)^2 > \frac{2}{3} + \frac{4}{9}$$

$$\left(x + \frac{2}{3}\right)^2 > \frac{10}{9}$$

Thus,

$$x + \frac{2}{3} > \frac{\sqrt{10}}{3} \qquad \text{or} \quad x + \frac{2}{3} < -\frac{\sqrt{10}}{3}$$

$$x > \frac{-2}{3} + \frac{\sqrt{10}}{3} \quad \text{or} \quad x < \frac{-2}{3} - \frac{\sqrt{10}}{3}$$

$$x > \frac{-2 + \sqrt{10}}{3} \quad \text{or} \quad x < \frac{-2 - \sqrt{10}}{3}$$

EXERCISES

1. Solve the following inequalities.
   (a) $x^2 > 49$ (b) $x^2 < 64$
   (c) $(x + 3)^2 < 49$ (d) $(x - 5)^2 > 64$
   (e) $(2x + 1)^2 > 49$ (f) $(3x - 1)^2 < 64$
   (g) $(5x + 7)^2 < 49$ (h) $(6x + 5)^2 < 64$
   (i) $(3x + 2)^2 > 0$ (j) $(4x - 1)^2 < 0$

2. Solve the following inequalities.
   (a) $x^2 < 36$ (b) $x^2 > 81$
   (c) $(x + 5)^2 > 36$ (d) $(x - 4)^2 < 81$
   (e) $(2x - 1)^2 > 36$ (f) $(3x + 1)^2 < 81$
   (g) $(7x + 5)^2 < 36$ (h) $(6x + 7)^2 > 81$
   (i) $(5x - 2)^2 > 0$ (j) $(6x - 7)^2 < 0$

3. Solve the following inequalities.
   (a) $x^2 > 13$ (b) $x^2 < 17$
   (c) $(x + 2)^2 < 13$ (d) $(x - 7)^2 > 17$
   (e) $(2x - 3)^2 < 13$ (f) $(3x + 2)^2 > 17$
   (g) $(5x + 6)^2 > 13$ (h) $(7x - 3)^2 < 17$
   (i) $(3x + 5)^2 > -7$ (j) $(2x - 1)^2 < -2$

4. Solve the following inequalities.
   (a) $x^2 < 19$ (b) $x^2 > 23$
   (c) $(x - 3)^2 > 19$ (d) $(x + 7)^2 < 23$
   (e) $(3x - 2)^2 > 19$ (f) $(5x + 1)^2 < 23$
   (g) $(7x - 4)^2 < 19$ (h) $(5x + 7)^2 > 23$
   (i) $(2x - 9)^2 > -4$ (j) $(6x + 11)^2 < -5$

5. Solve the following inequalities.
   (a) $x^2 + 6x > 8$ (b) $x^2 - 4x < 10$
   (c) $x^2 - 10x - 12 > 0$ (d) $x^2 + 8x + 3 < 0$
   (e) $2x^2 + 4x < 7$ (f) $3x^2 + 9x - 8 > 0$
   (g) $5x^2 + 2x > 3$ (h) $6x^2 + 2x - 5 < 0$
   (i) $3x^2 + 2x + 7 > 0$ (j) $5x^2 - 3x + 8 < 0$

6. Solve the following inequalities.
   (a) $x^2 - 6x > 12$ (b) $x^2 + 12x < 11$
   (c) $x^2 + 14x - 5 > 0$ (d) $x^2 + 4x + 1 < 0$
   (e) $2x^2 - 8x < 5$ (f) $3x^2 - 6x - 5 > 0$
   (g) $7x^2 - 3x > 2$ (h) $5x^2 + 4x - 7 < 0$
   (i) $4x^2 + 3x + 9 > 0$ (j) $2x^2 + 3x + 10 < 0$

7. Under the assumption that $A > 0$, solve the general inequality
   $Ax^2 + Bx + C > 0$

8. Under the assumption that $A > 0$, solve the general inequality
   $Ax^2 + Bx + C < 0$

9. If p is the average of the solutions of $Ax^2 + Bx + C = 0$, and $A > 0$, is $Ap^2 + Bp + C \leq 0$ or is $Ap^2 + Bp + C \geq 0$? Justify your answer.

10. (a) If $A > 0$, must the inequality $Ax^2 + Bx + C \geq 0$ have solutions? Justify your answer.
    (b) If $A > 0$, must the inequality $Ax^2 + Bx + C \leq 0$ have solutions? Justify your answer.

6. Applications

The problems that you considered in Chapter 2 could all be solved by using linear equations or inequalities. As you might expect, more complicated problem situations lead to more complicated equations and inequalities. In this section we consider some problems which can be solved using polynomial equations and inequalities. The technique is essentially the same as before:

A. Analyze the problem to determine the known relationships.
B. Determine an equation or inequality involving the quantities under consideration.
C. Represent these quantities algebraically.
D. Rewrite your equation or inequality using the representation.
E. Solve and translate your algebraic solutions into real answers.

Example 1

The sum of two numbers is 8 and their product is 10. Determine the numbers.

In this case we have a mathematical puzzle and the information almost jumps out at us. One of the equations is

(first number)(second number) $= 10$

The other is

(first number) + (second number) $= 8$

We could use two variables and two equations and then solve using substitution.

Let x be the first number and let y be the second number.

The two equations are

$xy = 10$ and $x + y = 8$

Solving the second for y yields

$y = 8 - x$

Substituting in the first equation yields

$$x(8 - x) = 10$$
$$8x - x^2 = 10$$
$$x^2 - 8x + 10 = 0$$

Using the quadratic formula yields

$$x = \frac{8 \pm \sqrt{64 - 40}}{2}$$

$$= \frac{8 \pm \sqrt{24}}{2}$$

$$= \frac{8 \pm 2\sqrt{6}}{2}$$

$$= 4 \pm \sqrt{6}$$

If $x = 4 + \sqrt{6}$, $y = 8 - (4 + \sqrt{6}) = 4 - \sqrt{6}$. If $x = 4 - \sqrt{6}$, $y = 8 - (4 - \sqrt{6}) = 4 + \sqrt{6}$. The two numbers are $4 + \sqrt{6}$ and $4 - \sqrt{6}$. $\square$

In this example, we used two variables and then eliminated one of them by substitution. We could have saved some work just by different representation. For example, if we let x be the first number, then $8 - x$ is the second number and we obtain the equation

$$x(8 - x) = 10$$

We then proceed as before.

Example 2

Using physics and calculus, it can be shown that if an object is thrown upward with a velocity of v_0 at time 0, then its distance d above the ground at time t is given approximately by

$$d = -16t^2 + v_0 t$$

The velocity must be given in feet per second and d is measured in feet. Realize that this equation holds only from the time that the object is thrown until it returns to the ground. (The equation also ignores the friction due to air.)

If a ball is thrown upward with an initial velocity of 64 ft/sec, at what time will it hit the ground?

In this case $v_0 = 64$ and we wish to know for what values of t is $d = 0$. Since $v_0 = 64$,

$$d = -16t^2 + 64t$$

We need to solve

$$0 = -16t^2 + 64t$$

This can be rewritten as

$$16t^2 - 64t = 0$$

Factoring yields

$$16t(t - 4) = 0$$

Either $t = 0$ or $t - 4 = 0$. Thus $t = 0$ or $t = 4$.

Since the object is thrown at time 0, it must return to the ground after 4 sec. □

Example 3

Among the more useful geometric facts is the Pythagorean theorem which states:

If a, b, and c are the three sides of a right triangle and c is the side opposite the right angle, then

$a^2 + b^2 = c^2$

Suppose a plane is flying north at an airspeed of 240 mi/hr and there is a wind blowing from west to east at 30 mi/hr. At the end of $\frac{1}{2}$ hr, how far will the plane have traveled? The situation is illustrated in Figure 5-6. We can view this as if the plane flew north for $\frac{1}{2}$ hr and then was blown eastward for $\frac{1}{2}$ hr. The northward distance then is 120 mi and the eastward distance is 15 mi. Using the Pythagorean theorem we have

$x^2 = (15)^2 + (120)^2$

Thus

$$x^2 = 225 + 14{,}400$$
$$= 14{,}625$$
$$x = \sqrt{14{,}625}$$
$$= \sqrt{25 \cdot 585}$$
$$= 5\sqrt{585}$$

Using a table, we see that

$x \doteq 121$ □

Figure 5-6

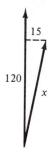

15

120

x

Example 4

Suppose that you have 100 ft of fence with which to enclose a rectangular garden. What shape will give you the maximum area?

Our intuition and experience might lead us to expect that a square would be the best shape. In order to solve this problem, we would have to demonstrate that any other rectangular shape yields a smaller area. If we let x denote one side of the rectangle, then x also denotes the opposite side. Since the perimeter is 100 ft, the other sides are each

$$(\tfrac{1}{2})(100 - 2x) = 50 - x$$

(See Figure 5-7.)

The area of the rectangle depends on x and is given by

$$A = x(50 - x)$$

If we choose the square shape, then

$$A = 25 \cdot 25 = 625$$

In order to satisfy ourselves that the square is the best choice we need to show that

$$A \leq 625$$

for all possible choices of x. In order to do so, we solve this inequality. The inequality

$$A \leq 625$$

is equivalent to

$$0 \leq 625 - A$$

Substituting yields

$$0 \leq 625 - [x(50 - x)]$$
$$0 \leq 625 - 50x + x^2$$
$$0 \leq (25 - x)^2$$

Figure 5-7

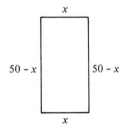

Since $(25 - x)^2 \geq 0$ for all values of x, $A \leq 625$ for all values of x and the area of the garden is never greater than 625. If $x \neq 25$, $(25 - x)^2 > 0$ and $A < 625$. Thus the square shape yields the maximum area. $\square$

EXERCISES

1. The product of two numbers is 20 and their difference is 4. What are the numbers?
2. If a gun is fired upward from ground level and the bullet leaves the gun with a velocity of 1280 ft/sec, how much time elapses before it returns to the ground?
3. Determine a decimal approximation for the length of the diagonal of a rectangle if the width is 5 and the length is 15.
4. One side of a rectangle is 5 ft longer than the other and the area is 150 sq ft. What are the dimensions?
5. The volume of a sphere is given by $V = \frac{4}{3}\pi r^3$ where r denotes the radius. Suppose you have two spheres, one with radius 2 in. and the other with unknown radius but suppose you know that the one with unknown radius has twice the volume of the other. What is its radius? If we double the radius of a sphere, how does that affect the volume?
6. Suppose you have 200 ft of fence and wish to enclose a rectangular garden with area of about 1000 sq ft. To the nearest foot, what should be the dimensions of the garden?
7. Show that a circular region enclosed by 100 ft of fence is larger than a square region enclosed by 100 ft of fence. (The circumference of a circle is given by $C = 2\pi r$ and the area is given by $A = \pi r^2$.)
8. In geometry it is shown that any triangle constructed with all vertices on a circle and with a diameter of the circle as one side is a right triangle, the right angle being opposite the diameter. (See Figure 5-8.)

    Furthermore, any rectangle constructed with all vertices on the circle consists of two such triangles. Suppose you have a circle with diameter 1. What is the largest rectangle that can be so constructed and what is its shape?

Figure 5-8

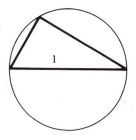

9. Suppose that experience shows that the average shipping cost C per unit for a particular machinery business is given by the equation $C = kd^2$ for some constant k where d is the average distance from the factory to the customer.
   (a) Find k if the average shipping cost in one year is $42 and the average distance from the factory to the customer is 100 mi.
   (b) What would you expect the average shipping cost to be the next year if the business is extended so that the average customer is 120 mi from the factory?

10. The strength of a wire or rope is given by the formula $S = kd^2$ where S is the breaking strength, d is the diameter, and k is a constant.
   (a) Determine k if a wire made of a particular alloy with a diameter of .025 in. breaks at 10 lb.
   (b) What is the breaking strength of a wire with diameter .05 in.?
   (c) What is the diameter of a wire with a breaking strength of 20 lb?

11. The intensity of light is given by the equation $I = c/d^2$ where I is the intensity, d is the distance from the source, and c is a constant.
   (a) Determine c if the intensity at a distance of 5 cm is 20 units.
   (b) What is the intensity at a distance of 10 cm?
   (c) If the intensity is 25 units, what is the distance from the source?

12. The force F of gravitational attraction between two objects with fixed masses is governed by the equation $F = k/d^2$ where d is the distance between the objects and k is a constant.
   (a) Find k if $F = 12$ when $d = 200$ cm.
   (b) What is the force when these same objects are 250 cm apart?
   (c) How far apart are they if the force is 20 units?

13. From experience it is found that the monthly profit P of a certain appliance business is given by the equation $P = .05S^2 - 7S$ where S denotes the number of units sold.
   (a) What are the profits on sales of 200 units?
   (b) What must the sales be in order to show some profit, that is, avoid a loss?
   (c) What sales are necessary for a profit of at least $1200 a month?

14. The distance d that a car with speed s will travel after the breaks are applied is given by the equation $d = .02s^2 + .7s$ where speed is given in feet per second and d is in feet.
   (a) If an automobile is going 60 mph, how far will it travel after the breaks are applied?
   (b) If skid marks measure 180 ft, how fast was the car going when the breaks were slammed on?
   (c) What speed provides a stopping distance of no more than 50 ft after the

15. Two boys are to mow a yard 80×100. Their procedure is to mow completely around the rectangle and keep going until the job is finished. How wide a swath should the first mow so that they mow the same amount?

16. Two private planes leave an airport, one flying south at 150 mph and the other flying east at 210 mph.
   (a) How far apart will they be in $\frac{1}{2}$ hr?
   (b) How long will it take them to be at least 540 mi apart?

17. Suppose that a rocket is launched with an initial velocity of 80 ft/sec.
   (a) How long will it take to reach a height of 40 ft?
   (b) Will it ever reach a height of 150 ft? Explain your answer.

18. Suppose your experience in the carwash business tells you that the number of cars washed per day and the price charged per car are related by the equation $N = 600 - 200p$ where N is the number washed, p is the price, and $0 \leq p \leq \$3$. Furthermore, suppose your costs are given by $C = 30 + .25N$.
    (a) Express the profit P in terms of the price p.
    (b) What will be the profits at a price of $1.50?
    (c) What should the price be in order to assure a profit of at least $75 a day?
    *(d) Could you charge more and make more? How much and how much?

6

Graphs

The French mathematician and philosopher Descartes invented analytic geometry and thus provided us with one of the most useful tools in mathematics. In this chapter we will see how algebra and geometry can be blended to use methods and concepts from one area to enlighten our understanding of the other.

1. The Coordinate System and the Concept of a Graph

Intuitively, we expect to be able to assign a real number to each point on a line so that the distance between two points will be the absolute value of the difference between the corresponding numbers or *coordinates*. We discussed number lines in Chapter 1 and in Chapters 2 and 5 we used them to describe solution sets of systems of inequalities. (See Figure 6-1.)

Figure 6-1

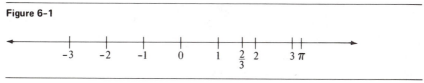

Just as we established a correspondence between numbers and points on a line, we now wish to establish a correspondence between *ordered pairs* of numbers and points in the plane and thereby establish a system of coordinates for the plane. In order to establish these coordinates we use two perpendicular number lines (usually we call one X and the other Y) and choose the coordinates so that the intersection of the lines has the coordinate 0 on both lines; that is, as a point on Y, the intersection is named by 0 and also as a point on X it is named by 0. It is customary to orient the lines as in Figure 6-2.

Figure 6–2

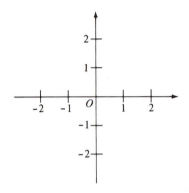

We could choose other orientations, but for most purposes the one above is sufficient. Now let P be any point in the plane. If we construct the lines through P parallel to Y and X, then these lines will pass through X and Y and, in fact, will be perpendicular to them. We then assign to P the pair of numbers (a, b), where a is the coordinate of the foot of the perpendicular from P to X and b is the coordinate of the foot of the perpendicular from P to Y. The numbers a and b are called the *coordinates* of P, and we use the ordered pair (a, b) to indicate that a is the "X-coordinate" and b the "Y-coordinate."

In Figure 6-3 the coordinates of P are $(2.1, 3.4)$. It is customary to speak or write of the "point (a, b)" rather than the "point P with coordinates (a, b)." If we are given an ordered pair of numbers, we can locate a point by reversing the above process. For example, the point $(5, -2)$ is the intersection of the vertical line through the point "5" on X and the horizontal line through the point "-2" on Y. The fundamental assumption of analytic geometry is that the above process determines a one-to-one correspondence between points in the plane and the set of all ordered pairs of real numbers. Such a correspondence is called a *coordinate system* for the plane.

Figure 6-3

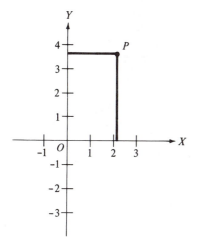

In plotting points, remember that the first number is the X-coordinate and the second is the Y-coordinate; for example, the order in the pair (4, 3) is the same as the alphabetical order of X and Y. Locating points on a coordinate system is easier if we use coordinate paper as in Figure 6-4.

Figure 6-4

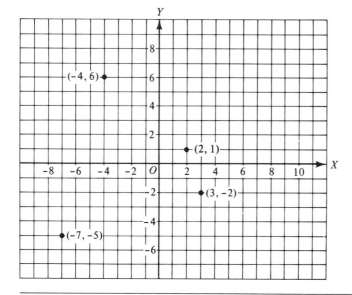

In Chapter 2, when we drew the graph of the solution set of an inequality or of a system of inequalities, we were just drawing a picture of a set of numbers. Now we wish to draw graphs or pictures of sets of ordered pairs of numbers. Our primary interest will be in drawing graphs of solution sets of equations and inequalities. In Chapter 2, when we discussed equations such as $2x + 5y = 7$, we said that the solution set was a set of ordered pairs of numbers. To graph such an equation means to draw a picture of all of these ordered pairs on a standard coordinate system. The general procedure is to compute several of the pairs, plot them, and from this preliminary information, try to determine just what the set looks like and complete the drawing. Frequently it is convenient to make a table of values of x and corresponding values of y rather than write down the pairs.

Example 1

Draw the graph of

$$y = 2x - 3$$

First we make a table for some of the solutions by substituting values for x and determining corresponding values for y. For example, $x = 0$ yields $y = -3$.

x	0	1	2	4	-1
y	-3	-1	1	5	-5

Figure 6–5

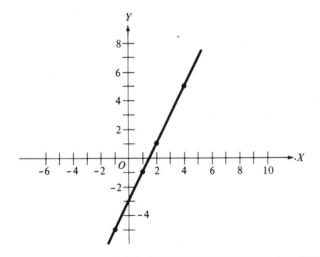

From the table we have the set of points

$$\{(0, -3), (1, -1), (2, 1), (4, 5), (-1, -5)\}$$

When we plot these on a coordinate system, it appears as if all the points lie on one straight line and we complete the graph by drawing the line through these points. See Figure 6-5. ☐

Example 2

Draw the graph of

$$x + 2y = 3$$

We make a table as before.

x	0	1	-1	2	-2	4	-4	5
y	$\frac{3}{2}$	1	2	$\frac{1}{2}$	$\frac{5}{2}$	$-\frac{1}{2}$	$\frac{7}{2}$	-1

Now we plot the points just by reading the coordinates from the table. (See Figure 6-6.) Again the points all seem to lie on one straight line and we draw the graph by drawing the straight line through these points. ☐

Figure 6-6

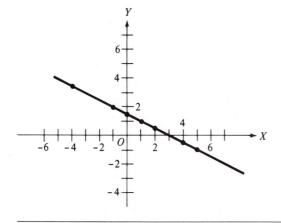

Example 3

Draw the graph of

$$y = x^2$$

x	0	1	−1	2	−2	3	−3
y	0	1	1	4	4	9	9

See Figure 6-7. In this case the points were not all on one straight line and we connected them using a smooth curve. □

Figure 6-7

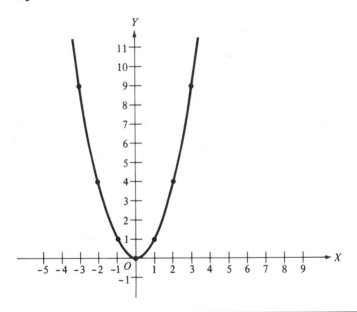

The general procedure illustrated above can be applied to any equation but it is cumbersome and time-consuming. The rest of this chapter is devoted to the development of techniques that will enable us to draw graphs for certain forms of equations with much more ease. Before proceeding further, however, you need practice with the basic "plot and sketch" method.

EXERCISES

1. Plot the following sequence of points and join each point to the next by a straight line segment.

   (2, 5), (5, 5), (5, 3), (2, 3), (2, 5), (7, 7), (10, 7), (10, 5), (5, 3), (5, 5), (10, 7)

2. Plot the following sequence of points and join each point to the next by a straight line segment.

$(-5, 6), (5, 6), (5, 15), (-5, 15), (-5, 6), (-2, 6), (-2, 10), (0, 11), (2, 13),$
$(1, 11), (-1, 9), (-2, 6), (0, 9), (2, 11), (3, 13), (3, 11), (1, 6), (3, 8), (2, 6),$
$(5, 6), (5, -3), (-2, -3), (-2, 5), (1, 5), (1, -2), (0, -3), (-5, -3), (-5, 6)$

Draw the graphs of the following equations.

3. $y = 2x + 1$	4. $y = -3x + 4$
5. $2x + y = 7$	6. $2x + 3y = 6$
7. $y = x^2 + 1$	8. $y = x^2 - 3$
9. $y = 2x^2$	10. $y = -3x^2$
11. $y = \frac{1}{4}x^2$	12. $y = x^2 + 2x + 1$
13. $y = x^2 - 4x + 4$	14. $y = x^2 - 4x + 6$
15. $y^2 = x$	16. $y = 5$
17. $x = 2$	18. $y = x^3$
19. $y = x^3 + 1$	20. $y = x^3 - 2$
21. $y^3 = x$	22. $y = x^4$
23. $y = x^5$	24. $y = 1/x$
25. $y = 1/(x - 1)$	26. $y = 1/(x + 2)$
27. $y = x$	28. $y^2 = x^2$
*29. $|y| = |x|$	*30. $x^2 + y^2 = 16$
*31. $x^2 + y^2 = 1$	*32. $x^2 - y^2 = 1$

2. Lines

In geometry the word "line" means the same as "straight line" and we will follow this custom. Thus all lines under discussion are "straight lines."

In some of the examples and exercises for Section 1, the graph was a line. Using the concept of similarity from geometry, it can be shown that every line is the graph of an equation of the form

$$px + qy = c$$

where p, q, and c are numbers with $p \neq 0$ or $q \neq 0$. Furthermore, the graph of any such equation is a line.

We know that two points completely determine a line, that is, given two points, there is exactly one line through both of them. If we have an equation whose graph is a line, we can compute two pairs in the solution set, plot the points, and complete the graph just by drawing the line through the points. We get a better graph if we pick points that are not too close together.

Example 1

Draw the graph of

$$2x + 5y = 3$$

Since this equation has a graph which is a line, we only need to determine two points in the solution set.

If $x = -1$, $y = 1$ and therefore $(-1, 1)$ is on the line.
If $x = 4$, $y = -1$ and therefore $(4, -1)$ is on the line.

We plot $(-1, 1)$ and $(4, -1)$ and draw the line joining them as in Figure 6-8. ☐

Figure 6–8

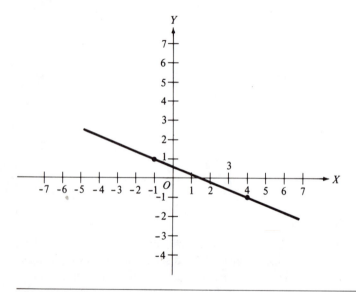

The equations of horizontal and vertical lines are somewhat special but, in fact, are the simplest. The horizontal line through the point $(2, 5)$ must be 5 units above the X-axis everywhere and therefore each point on the line must have Y-coordinate 5. Thus the equation of this line is just $y = 5$. The vertical line through this same point is the set of all points with X-coordinate 2 and thus its equation is just $x = 2$. (The graphs are given in Figure 6-9.) Since the general equation of a line is of the form

$$px + qy = c$$

if $p \neq 0$ and $q \neq 0$, the points $(0, c/q)$ and $(c/p, 0)$ are both on the line and

Figure 6-9

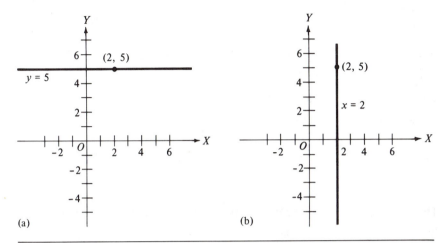

(a)   (b)

it is neither horizontal nor vertical. If $p = 0$, then the equation becomes $qy = c$ or $y = c/q$ and the graph is the horizontal line through $(0, c/q)$. If $q = 0$, the equation becomes $px = c$ or $x = c/p$ and the graph is the vertical line through $(c/p, 0)$.

Example 2

The graph of $5x = 6$ is the vertical line through $(\frac{6}{5}, 0)$.
The graph of $2y = 6$ is the horizontal line through $(0, 3)$. $\square$

Suppose two equations of the form

$$y = mx + b \quad \text{and} \quad y = px + q$$

have the same graph. Since the coefficient of y is not zero, the line is not vertical and meets the Y-axis (the line with equation $x = 0$) at exactly one point. According to the first equation, this is the point $(0, b)$. According to the second equation, this point is the point $(0, q)$. Thus $(0, b) = (0, q)$ and $b = q$. Furthermore, the line meets the vertical line with equation $x = 1$ in exactly one point. By the first equation this is the point $(1, m + b)$. By the second it is $(1, p + q)$. Thus $m + b = p + q$. But we already know that $b = q$. Thus $m + b = p + b$ and $m = p$. Thus the two equations are identical and we see that *two equations of the form*

$$y = mx + b$$

have the same graph if and only if the equations are identical.

If we solve the equation

$$2x + 5y = 7$$

for y in terms of x we have

$$5y = -2x + 7$$
$$y = -\tfrac{2}{5}x + \tfrac{7}{5}$$

In general, if $q \neq 0$, the equation

$$px + qy = c$$

can be solved for y in terms of x.

$$px + qy = c$$
$$qy = -px + c$$
$$y = -\frac{p}{q}x + \frac{c}{q}$$

Thus if a line is not vertical, it is the graph of exactly one equation of the form

$$y = mx + b$$

This form of the equation for a line is called the *slope-intercept* form; the number m is called the *slope* of the line and b is called the *Y-intercept*. Since a vertical line does not have an equation of this form, *vertical lines have no slope*.

In our previous discussion we saw that $(0, b)$ is the point where the line meets the Y-axis, that is, the point where the line *intercepts* the Y-axis. This explains the name for b.

The motivation for the word slope is better understood by looking at the graphs given in Figure 6-10.

From these examples it seems that the larger the absolute value of m, the "steeper" the line and that lines with positive slope "run uphill" to the right while lines with negative slope "run downhill to the right." This is, in fact, true in general.

If we have an equation in slope-intercept form, we can use the slope and intercept to plot two points quite easily and then just draw the line. Using the equation

$$y = mx + b$$

we know $(0, b)$ is on the line. Furthermore, $(1, m + b)$ is on the line and we now can draw the graph using those two points. Notice as x changes from 0 to 1, y changes from b to $m + b$. Thus as x changes from 0 to 1, the change in y is just m.

If m is positive, then $m + b > b$. Thus $(1, m + b)$ is "higher" than $(0, b)$, and the line "goes uphill to the right." If m is negative, then $m + b < b$. Thus $(1, m + b)$ is "lower" than $(0, b)$ and the line "goes downhill to the right" as illustrated in Figure 6-11.

Figure 6–10

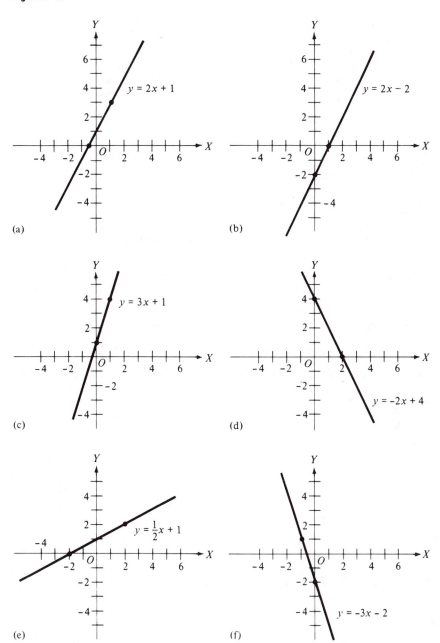

(a)

(b)

(c)

(d)

(e)

(f)

Figure 6-11

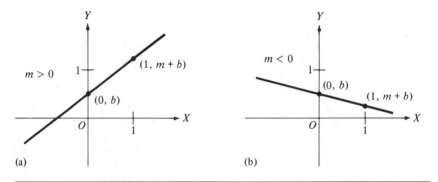

Example 3

The graph of $y = 3x + 2$ is given in Figure 6-12(a). ☐

Example 4

The graph of $y = -2x - 1$ is given in Figure 6-12(b). ☐

Suppose we have two equations

$$y = m_1 x + b \quad \text{and} \quad y = m_2 x + b$$

with $m_1 > m_2 > 0$. Then the point $(0, b)$ is on both lines but $(1, m_1 + b)$ is

Figure 6-12

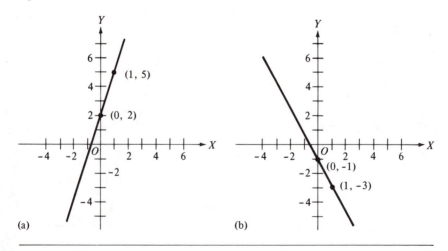

on the first line whereas $(1, m_2 + b)$ is on the second line. Since $m_1 > m_2$, $(1, m_1 + b)$ is above $(1, m_2 + b)$ and the first line is steeper as illustrated by Figure 6-13.

If $m_1 < m_2 < 0$, then $(1, m_1 + b)$ is below $(1, m_2 + b)$ and the lines are running downhill to the right. Thus the first line is steeper than the second as in Figure 6-14.

Figure 6–13

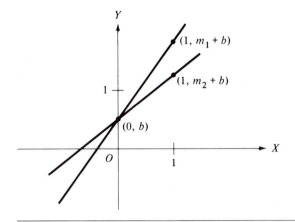

Figure 6–14

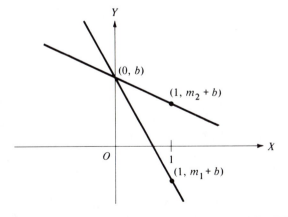

EXERCISES

In exercises 1–20, draw graphs of the given equation.

1. $x = \frac{5}{2}$
2. $y = -\frac{3}{4}$
3. $5x = -13$
4. $y = 0$
5. $y = 3x - 4$
6. $y = 2x + 5$
7. $y = \frac{1}{3}x + 2$
8. $y = -4x + 3$
9. $y = -\frac{1}{2}x + 4$
10. $y = 6x - 5$
11. $4x + 2y = 5$
12. $x + 3y = 6$
13. $2x = 5y = 1$
14. $6x + 3y + 5 = 0$
15. $7x + 5y = 6$
16. $y = \sqrt{2}x + 1$
17. $y = 2x - \sqrt{3}$
18. $y = \sqrt{5}x + \sqrt[3]{7}$
19. $\sqrt{2}x + \sqrt{3}y = 1$
20. $\sqrt{5}x - 2y = 3$

21. (i) Compute at least four points on the graph of $3x + 2y = 5$. Label these points P_1, P_2, P_3, and P_4, and let $P_1 = (x_1, y_1)$, $P_2 = (x_2, y_2)$, $P_3 = (x_3, y_3)$, and $P_4 = (x_4, y_4)$. For example, you might have $P_1 = (-1, 4)$ and then $x_1 = -1$ and $y_1 = 4$. Now compute the following numbers.

(a) $\dfrac{y_2 - y_1}{x_2 - x_1}$ (b) $\dfrac{y_4 - y_2}{x_4 - x_2}$ (c) $\dfrac{y_3 - y_1}{x_3 - x_1}$ (d) $\dfrac{y_4 - y_3}{x_4 - x_3}$

(e) $\dfrac{y_4 - y_1}{x_4 - x_1}$ (f) $\dfrac{y_1 - y_2}{x_1 - x_2}$ (g) $\dfrac{y_3 - y_2}{x_3 - x_2}$ (h) $\dfrac{y_1 - y_3}{x_1 - x_3}$

(ii) Transform the equation $3x + 2y = 5$ into an equation in slope-intercept form. What is the slope?

22. Repeat Exercise 21 using the equation $4x + 5y = 6$.

23. On the basis of Exercises 21 and 22, would you like to make a conjecture about the slope of a line?

24. If (x_1, y_1) and (x_2, y_2) are points on the graph of $y = mx + b$ with $x_2 \neq x_1$, what is

$$\frac{y_2 - y_1}{x_2 - x_1}?$$

25. Show that if $x_1 \neq x_2$, then

$$\frac{y_1 - y_2}{x_1 - x_2} = \frac{y_2 - y_1}{x_2 - x_1}$$

for any numbers x_1, x_2, y_1, y_2.

3. Determining Equations of Lines

Since a line is completely determined by any two points on it, any two points determine an equation for the line. In this section we develop a method of obtaining such an equation. Before developing any general techniques or formulas, let us consider some examples.

Example 1

Find an equation of the line through (2, 3) and (4, 8).

Since the line passes through two points with different X-coordinates, it is not vertical. Thus it has a slope and one of the general forms for the equation is

$$y = mx + b$$

Since (4, 8) and (2, 3) are on the line we must have

$$8 = 4m + b$$

and

$$3 = 2m + b$$

Thus

$$5 = 2m$$

and

$$m = \tfrac{5}{2}$$

Thus the slope of the line is $\tfrac{5}{2}$ and the equation is $y = \tfrac{5}{2}x + b$. We now need to determine b. Since (2, 3) is on the line we have

$$3 = \left(\tfrac{5}{2}\right) \cdot 2 + b$$

that is,

$$3 = 5 + b$$

$$b = -2$$

Thus

$$y = \tfrac{5}{2}x - 2$$

is an equation for the line. $\square$

Exercise 3 of Chapter 2, Section 6 is a problem much like this. By solving the general problem we now develop a formula for determining the slope of a line given two points on the line and obtain a general method for writing an equation for a line if we know the slope and a point or if we know two points.

Suppose two distinct points (x_1, y_1) and (x_2, y_2) are on the line with the equation $y = mx + b$.

$$y_1 = mx_1 + b \quad \text{and} \quad y_2 = mx_2 + b$$

Thus

$$y_1 - y_2 = (mx_1 + b) - (mx_2 + b)$$

$$y_1 - y_2 = (mx_1 - mx_2) + (b - b)$$
$$y_1 - y_2 = mx_1 - mx_2$$
$$y_1 - y_2 = m(x_1 - x_2)$$

Since the line is not vertical, $x_1 - x_2 \neq 0$, and therefore

$$m = \frac{y_1 - y_2}{x_1 - x_2}$$

(Look again at your answers to Exercises 21–25 in Section 2.) Note also that

$$m = \frac{y_2 - y_1}{x_2 - x_1}$$

Thus the slope of a nonvertical line is determined by any two distinct points on the line. *It is just the difference in Y-coordinates divided by the difference in X-coordinates*, that is, the change in Y-coordinates divided by the change in X-coordinates. This is sometimes called "the rise divided by the run," with $y_1 - y_2$ called the "rise" and $x_1 - x_2$ called the "run." (See Figure 6-15.) In the case of positive slope, the changes have the same sign and for negative slope they differ in sign. If the slope is 0, all Y-coordinates are the same, $y = b$, and the line is just the horizontal line through $(0, b)$.

Figure 6–15

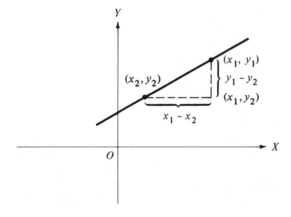

Example 2

The slope of the line through (4, 5) and (2, 1) is

$$\frac{5 - 1}{4 - 2} = \frac{4}{2} = 2 \quad \square$$

Example 3

The slope of the line through $(5, 2)$ and $(8, -3)$ is

$$\frac{2 - (-3)}{5 - 8} = \frac{5}{-3} = \frac{-5}{3} \ \square$$

Suppose we have a line with slope m and a point (x_1, y_1) on the line. If (x, y) is any other point on the line, then, since the line is not vertical, by our previous discussion, we know

$$\frac{y - y_1}{x - x_1} = m$$

Multiplying both sides by $(x - x_1)$ yields

$$y - y_1 = m(x - x_1)$$

Since every point on the line except (x_1, y_1) satisfies both equations and (x_1, y_1) satisfies the second, every point on the line satisfies the equation

$$y - y_1 = m(x - x_1)$$

This form is called a *point-slope* form of the equation of a line because it is immediately determined by the slope of a line and any point on the line.

If (x_1, y_1) and (x_2, y_2) are distinct points on a nonvertical line, then

$$\frac{y_1 - y_2}{x_1 - x_2} = m$$

is the slope of the line and

$$y - y_1 = \frac{y_1 - y_2}{x_1 - x_2}(x - x_1)$$

is an equation for the line. This is sometimes called the *two-point form* of the equation of a line.

We can use the point-slope form or the two-point form of the equation of a line to write down equations for lines if we know the slope and one point or if we know two points on the line.

Example 4

Give an equation for the line with slope 3 through $(5, 9)$.

We use the point-slope form and obtain

$$y - 9 = 3(x - 5) \ \square$$

Example 5

An equation for the line with slope -2 through $(1, -4)$ is given by

$$y - (-4) = -2(x - 1)$$

that is,

$$y + 4 = -2(x - 1) \;\square$$

Example 6

An equation for the line through $(6, 11)$ and $(4, 3)$ is given by

$$y - 3 = \frac{11 - 3}{6 - 4}(x - 4)$$

or

$$y - 3 = 4(x - 4) \;\square$$

Example 7

An equation for the line through $(7, -6)$ and $(5, -1)$ is given by

$$y - (-1) = \frac{(-6) - (-1)}{7 - 5}(x - 5)$$

or

$$y + 1 = -\tfrac{5}{2}(x - 5) \;\square$$

In these examples we could simplify the equations and put them in slope-intercept form but it is not always necessary or even desirable to do so.

Suppose we have two different lines having the same slope with equations

$$y = mx + b \quad \text{and} \quad y = mx + d$$

If the two lines meet at some point, say for $x = a$, then it follows that

$$(a, ma + b) = (a, ma + d)$$

Thus

$$ma + b = ma + d$$
$$b = d$$

and the lines must be the same because their equations are the same.

Thus if two *different* lines have the same slope, they are parallel. Conversely, if

$$y = mx + b \quad \text{and} \quad y = nx + d$$

are equations of two lines with $m \neq n$, then the point

$$\left(\frac{d - b}{m - n}, \frac{md - nb}{m - n} \right)$$

is on both lines and they meet. (We leave it to the reader to verify that this point is on both lines. Just substitute

$$\frac{d - b}{m - n}$$

for x in both equations.) Thus *two lines are parallel if and only if they have the same slope.*

EXERCISES

In exercises 1–36 find the equation of the line.

1. through $(0, 3)$ with slope 5
2. through $(0, 5)$ with slope 3
3. through $(5, 2)$ with slope 7
4. through $(1, 5)$ with slope 10
5. through $(2, 9)$ with slope -8
6. through $(5, 7)$ with slope -13
7. through $(-3, 4)$ with slope 4
8. through $(-6, 9)$ with slope 6
9. through $(10, 13)$ with slope $\frac{2}{3}$
10. through $(8, 11)$ with slope $\frac{5}{4}$
11. through $(\frac{1}{3}, \frac{2}{7})$ with slope 15
12. through $(\frac{4}{9}, \frac{1}{5})$ with slope 18
13. through $(7, 1)$ and $(2, 3)$
14. through $(6, 1)$ and $(8, 5)$
15. through $(0, 4)$ and $(1, 8)$
16. through $(0, 9)$ and $(2, 4)$
17. through $(5, -2)$ and $(8, 7)$
18. through $(-6, 5)$ and $(2, 9)$
19. through $(-4, 19)$ and $(17, 13)$
20. through $(16, 5)$ and $(-29, 14)$
21. through $(\frac{1}{2}, -\frac{1}{3})$ and $(\frac{1}{5}, \frac{1}{6})$
22. through $(\frac{1}{4}, \frac{1}{5})$ and $(\frac{1}{3}, -\frac{3}{4})$
23. through $(.13, 4.1)$ and $(.87, 5.3)$
24. through $(5.9, .17)$ and $(3.3, .51)$
25. through $(-.6, .4)$ and $(.7, -5)$
26. through $(.9, -.7)$ and $(-.4, .4)$
27. through $(4, 7)$ parallel to $x = 5$
28. through $(5, 9)$ parallel to $x = -4$
29. through $(1, 5)$ parallel to the line with equation $y = 5x - 4$
30. through $(-2, 4)$ parallel to the line with equation $y = 6x + 5$
31. through $(2, 3)$ parallel to the line with equation $y = -\frac{3}{2}x + 1$

32. through $(5, 7)$ parallel to the line with equation $y = -\frac{4}{3}x + 4$
33. the horizontal line through $(7, 3)$
34. the horizontal line through $(9, -12)$
35. the vertical line through $(7, 3)$
36. the vertical line through $(9, -12)$
37. If $m \neq n$, verify that

$$\left(\frac{d - b}{m - n}, \frac{md - nb}{m - n}\right)$$

satisfies both $y = mx + b$ and $y = nx + d$.

4. Linear Variation

Frequently two quantities are found to vary linearly with respect to each other, that is, they are related by a linear equation, one whose graph is a line. When this is the case, we need only to know two points on the line or two sets of associated values and we can determine the equation.

Example 1

Suppose you operate a small bakery with the capacity to bake 1000 loaves of bread a day, and suppose that the number that you sell varies linearly with the price per loaf. Let y be the number of loaves sold and x the price per loaf. Suppose, by past experience, you found that 69¢ is the highest price you can charge and still sell some bread. Then at 70¢ you will sell 0 loaves and, therefore, $(.70, 0)$ is a point on the line. On the other hand, you know that if you give the bread away you could get rid of over 1000 loaves. Suppose that, again by past experience, you know that 30¢ is the highest price at which you can sell 1000 loaves a day, that is, at 31¢ you will sell less than 1000 loaves. Then $(.30, 1000)$ is also a point on the line. Now, given two points we can also determine the slope of the line and, in this case, the slope is

$$\frac{1000 - 0}{.30 - .70} = \frac{1000}{-.40} = -2500$$

Using the point-slope form we now have

$$y = -2500(x - .70)$$

or

$$y = -2500x + 1750$$

This equation can now be used to compute the number of loaves that would be sold at any price between 30¢ and 70¢. For example, at 50¢,

$$y = -2500(.50) + 1750$$
$$= -1250 + 1750$$
$$= 500$$

Thus you will sell 500 loaves at 50¢ per loaf. At 60¢ per loaf we have

$$y = -2500(.60) + 1750$$
$$= -1500 + 1750$$
$$= 250 \quad \square$$

Example 2

It might seem reasonable to expect the gasoline mileage yielded by a particular model of an automobile engine to vary linearly with the weight of the vehicle it is used to power provided, of course, that gear rates, tire sizes, and so on, are the same. Thus, if we know the mileage figure for two cars of different weights with this same engine, we could predict the performance of this engine in other cars.

Suppose the engine yields 14.8 mpg in a car weighing 3900 lb and 13.2 mpg in a car weighing 4380 lb. If we let y denote mileage and x denote weight we have (3900, 14.8) and (4380, 13.2) on our line.

The slope of the line is given by

$$\frac{14.8 - 13.2}{3900 - 4380} = \frac{1.6}{-480} = -\frac{1}{300}$$

Using the point-slope form we have

$$y - 14.8 = -\frac{1}{300}(x - 3900)$$

$$y - 14.8 = -\frac{1}{300}x + 13$$

$$y = -\frac{1}{300}x + 27.8$$

Thus, if we mount this engine in an automobile weighing 3000 lb we have

$$y = -\frac{1}{300}(3000) + 27.8$$

$$= -10 + 27.8$$

$$= 17.8$$

and the mileage is 17.8 mpg.

If we mount this engine in an automobile weighing 4800 lb we have

$$y = -\frac{1}{300}(4800) + 27.8 = -16 + 27.8 = 11.8 \text{ mpg} \quad \square$$

Suppose y and x vary linearly and the line passes through the origin. The slope-intercept form of the equation is then

$$y = mx$$

In such a case, we say y *varies directly* with x. Thus *direct variation* is just a special case of linear variation. In fact, the statement that $y - b$ varies directly with x means that y varies linearly with x and the y-intercept of the line is just b.

EXERCISES

1. Suppose that y varies linearly with x, $y = 7$ when $x = 2$, and $y = -5$ when $x = 4$.
   (a) Give an equation expressing the relationship between x and y.
   (b) If $x = 8$, what is y?
2. Suppose s varies linearly with t, $s = 1.9$ when $t = 6$, and $s = 4.7$ when $t = 8$.
   (a) Give an equation expressing the relationship between s and t.
   (b) If $t = 13$, what is s?
3. Suppose that you are in the ice cream business and, from past experience, you believe that the amount sold varies linearly with the price charged. Furthermore, suppose you can sell 2000 gallons a day at $1.00 a gallon and 500 gallons a day at $2.25 a gallon.
   (a) If y denotes the number of gallons sold and x denotes the price charged, determine an equation which gives the relationship between x and y.
   (b) How many gallons a day would you sell at $1.50 a gallon?
   (c) How many gallons a day would you sell at $2.00 a gallon?
4. Suppose that from experiments it has been determined that the number of calories needed per day to maintain weight varies linearly with the weight of the individual, provided that the amount of activity remains constant.
   (a) If we let y denote the number of calories and x the weight, determine the relationship between x and y if a 140-lb man needs 2200 calories per day and a 200-lb man needs 2850.
   (b) How many calories does a 175-lb man need to maintain weight?
   (c) How many calories does a 155-lb man need?
   (d) Check the assumption for this problem by consulting a calorie table, choosing data from it instead of that given in part (a), and, after you determine the equation, compute your own table for the weights given in the table you are using.
5. Suppose that the number of no-return bottles of soft drinks sold varies directly with the price per bottle. It seems reasonable to assume further that if the price of the no-return bottle is the same as the price of the deposit bottle, then within a short period of time, none of the no-return bottles will be sold. For example, if no-return bottles cost the housewife 5¢ each and the deposit on regular bottles

is 5¢ each, she will not buy the no-return bottles. Let us assume that the deposit on regular bottles is 5¢ each and that when soft drinks in no-return bottles sell for 1¢ more than those in deposit bottles, that is, the no-return bottle costs the buyer 1¢, that drinks in no-return bottles claim 50 percent of the market.

(a) If y denotes the percentage of the market claimed by drinks in no-return bottles and x denotes the price of no-return bottles, what is the relationship between x and y?

(b) If the government puts a tax of 2¢ each on no-return bottles, how does that affect the number sold? A tax of 3¢? Of 4¢?

6. It seems reasonable to assume that the amount of salt needed by a county for snow removal will vary linearly with the amount of snow that falls each winter. It is not quite accurate to say that it varies directly because a new supply must be purchased each fall and even if no snow fell, the stockpile would deteriorate due to rain and other weather and have to be replaced or replenished the next year.

(a) If we let y be the number of tons of salt purchased and x be the inches of snow that fell, determine the relationship between x and y if the county used 70 tons in 1971–1972 when 80 in. of snow fell and 64 tons in 1972–1973 when 85 in. fell.

(b) How many tons would be needed if 92 in. of snow fell in 1973–1974?

(c) If they had 40 tons in stock on December 1 and 20 in. fell in December, how many tons should they buy to replenish the stock on January 1?

7. In certain businesses it is reasonable to expect profits to vary linearly with sales.

(a) If y denotes profits and x denotes sales, what is the relationship between profits and sales if a business made a profit of $33,000 on sales of $500,000 in 1972 and a profit of $37,000 on sales of $532,000 in 1973?

(b) What sales are necessary for the company to break even, that is, make zero profit, but not suffer a loss?

(d) If sales were $150,000, what would be the profits?

5. The Distance Formula, Circles

Since the coordinates of a point are determined by the distances from the point to the axes, we might expect the distance between two points to be determined by their coordinates. This is, in fact, the case. Before developing the general formula, let us consider a special case.

Example 1

Determine the distance between (3, 2) and (7, 5).

In Figure 6-16, we see that the points (3, 2), (7, 5), and (7, 2) form a right triangle with the hypotenuse being the line segment from (3, 2) to (7, 5). The two sides have lengths 3 and 4 and from the Pythagorean theorem we know that the length of the third side is given by

$$\sqrt{3^2 + 4^2} = \sqrt{9 + 16} = \sqrt{25} = 5 \quad \square$$

Figure 6-16

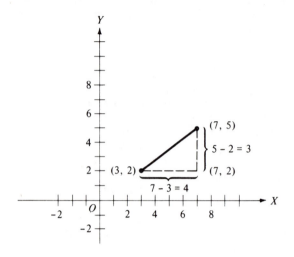

Figure 6-17

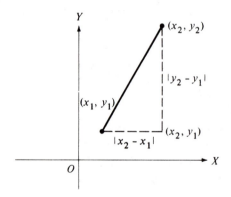

The above procedure can be generalized to provide a general formula for the distance between two points (x_1, y_1) and (x_2, y_2). The drawing in Figure 6-17 shows $x_1 < x_2$ and $y_1 < y_2$ but the argument does not depend on it.

The points (x_1, y_1), (x_2, y_2), and (x_2, y_1) form a right triangle with right angle at (x_2, y_1). The sides have length $|y_2 - y_1|$ and $|x_2 - x_1|$ and, by the Pythagorean theorem, the hypotenuse must have length given by

$$\sqrt{|x_2 - x_1|^2 + |y_2 - y_1|^2}$$

Since

$$|x_2 - x_1|^2 = (x_2 - x_1)^2$$

and

$$|y_2 - y_1|^2 = (y_2 - y_1)^2$$

the distance between the points is given by the *distance formula:*

$$[\text{distance from } (x_1, y_1) \text{ to } (x_2, y_2)] = \sqrt{(x_2 - x_1)^2 + (y_2 - y_1)^2}$$

Example 2

The distance from $(5, 2)$ to $(7, 9)$ is

$$\sqrt{(7 - 5)^2 + (9 - 2)^2} = \sqrt{2^2 + 7^2} = \sqrt{4 + 49} = \sqrt{53} \ \square$$

Example 3

The distance from $(4, 3)$ to $(-5, 5)$ is given by

$$\sqrt{(-5 - 4)^2 + (5 - 3)^2} = \sqrt{(-9)^2 + 2^2} = \sqrt{81 + 4} = \sqrt{85} \ \square$$

Example 4

The distance from $(17, 1)$ to $(13, 8)$ is given by

$$\sqrt{(17 - 13)^2 + (1 - 8)^2} = \sqrt{4^2 + 7^2} = \sqrt{16 + 49} = \sqrt{65}$$

Notice that it makes no difference which point is first. $\square$

We can use the distance formula to determine equations of geometric figures that are defined in terms of distances.

Example 5

Find an equation whose graph is a circle of radius 5 with center $(0, 0)$. This circle is

$$\{(x, y): \text{the distance from } (x, y) \text{ to } (0, 0) \text{ is } 5\}$$

By the formula this is

$$\{(x, y): \sqrt{(x - 0)^2 + (y - 0)^2} = 5\}$$
$$\{(x, y): \sqrt{x^2 + y^2} = 5\}$$

or

$$\{(x, y): x^2 + y^2 = 25\}$$

Thus $x^2 + y^2 = 25$ is an equation whose graph is the circle with radius 5 and center at $(0, 0)$. See Figure 6-18. $\square$

Figure 6-18

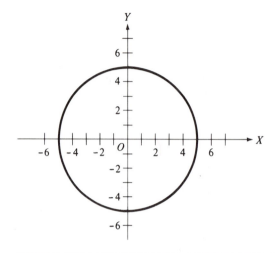

Let us generalize this result. The circle of radius r with its center at $(0, 0)$ is the set of points whose distance from $(0, 0)$ is r, that is, it is

$$\{(x, y): \text{the distance from } (x, y) \text{ to } (0, 0) \text{ is } r\}$$
$$= \{(x, y): \sqrt{(x - 0)^2 + (y - 0)^2} = r\}$$
$$= \{(x, y): \sqrt{x^2 + y^2} = r\}$$
$$= \{(x, y): x^2 + y^2 = r^2\}$$

Thus $x^2 + y^2 = r^2$ is an equation whose graph is a circle of radius r with center at $(0, 0)$.

Example 6

The graph of $x^2 + y^2 = 9$ is given in Figure 6-19(a). $\square$

Example 7

The graph of $x^2 + y^2 = \dfrac{81}{4}$ is given in Figure 6-19(b). $\square$

Example 8

What is the equation of a circle with center $(4, 1)$ and radius 5? This is the set

$$\{(x, y): \sqrt{(x - 4)^2 + (y - 1)^2} = 5\} = \{(x, y): (x - 4)^2 + (y - 1)^2 = 25\}$$

Figure 6-19

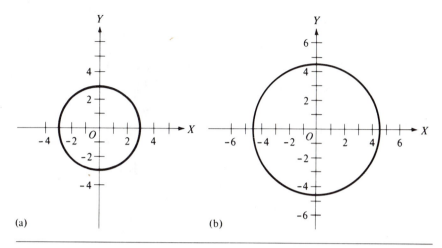

(a)   (b)

Thus $(x - 4)^2 + (y - 1)^2 = 25$ is the desired equation. See Figure 6-20.

We can generalize to a circle of radius r with center at the point (h, k). This circle is

$\{(x, y): \text{the distance from } (x, y) \text{ to } (h, k) \text{ is } r\}$
$= \{(x, y): \sqrt{(x - h)^2 + (y - k)^2} = r\}$
$= \{(x, y): (x - h)^2 + (y - k)^2 = r^2\}$

Figure 6-20

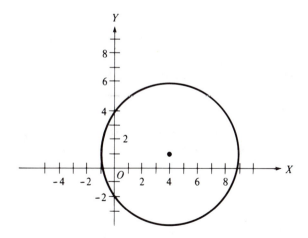

Figure 6–21

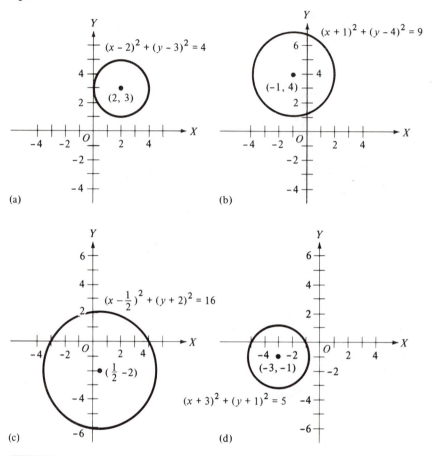

(a)   (b)   (c)   (d)

Thus

$$(x - h)^2 + (y - k)^2 = r^2$$

is an equation for the circle of radius r with center (h, k).

To graph such an equation we need only to locate the center, determine the radius, and use a compass. See Figure 6-21. ☐

Sometimes an equation does not fall immediately into this pattern but can be transformed into one of the desired form.

Example 9

$$x^2 - 2x + y^2 - 6y = 6$$
$$x^2 - 2x + 1 + y^2 - 6y + 9 = 6 + 1 + 9$$
$$(x - 1)^2 + (y - 3)^2 = 16 \quad \square$$

The graph of this equation is a circle with center $(1, 3)$ and radius 4. In transforming the equation, we used the "completing the square technique" twice in that we added suitable numbers to both sides in order to make the left side the sum of two perfect squares of the desired form.

Example 10

$$x^2 - 4x + y^2 - 10y = 7$$
$$x^2 - 4x + 4 + y^2 - 10y + 25 = 7 + 4 + 25$$
$$(x - 2)^2 + (y - 5)^2 = 36$$

The graph is a circle with center $(2, 5)$ and radius 6 as in Figure 6-22(a). □

Figure 6-22

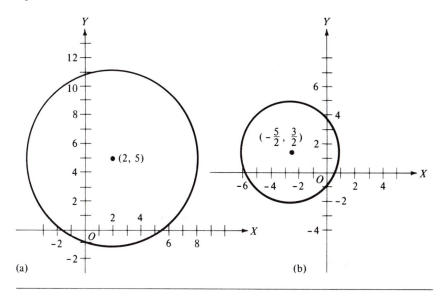

(a)

(b)

Example 11

$$x^2 + 5x + y^2 - 3y = 4$$
$$x^2 + 5x + \frac{25}{4} + y^2 - 3y + \frac{9}{4} = 4 + \frac{25}{4} + \frac{9}{4}$$
$$\left(x + \frac{5}{2}\right)^2 + \left(y - \frac{3}{2}\right)^2 = \frac{50}{4}$$

The graph is a circle with center $(-5/2, 3/2)$ and radius $5\sqrt{2}/2$ as in Figure 6-22(b). □

EXERCISES

1. Compute the distances between the following pairs of points.
   (a) $(4, 7)$ and $(3, 2)$ (b) $(4, -1)$ and $(7, 3)$
   (c) $(5, 2)$ and $(8, -6)$ (d) $(4, 11)$ and $(7, 19)$
   (e) $(5, 6)$ and $(6, 5)$ (f) $(1, 3)$ and $(-1, -3)$
   (g) (a, b) and $(-a, -b)$ (h) (a, b) and $(a\sqrt{2}, b\sqrt{2})$

2. Compute the distances between the following pairs of points.
   (a) $(5, 9)$ and $(4, 2)$ (b) $(6, -2)$ and $(8, 3)$
   (c) $(9, 6)$ and $(12, -4)$ (d) $(5, 13)$ and $(7, 19)$
   (e) $(7, 8)$ and $(8, 7)$ (f) $(2, 5)$ and $(-2, -5)$
   (g) (p, q) and $(-p, -q)$ (h) (p, q) and $(p\sqrt{2}, q\sqrt{2})$

3. Draw graphs of the following equations.
   (a) $x^2 + y^2 = 16$ (b) $x^2 + y^2 = \frac{9}{4}$
   (c) $(x - 1)^2 + (y - 2)^2 = 16$ (d) $(x + 2)^2 + (y - 3)^2 = 16$
   (e) $(x - 2)^2 + (y - 5)^2 = \frac{9}{4}$ (f) $(x + \frac{1}{2})^2 + (y - \frac{1}{4})^2 = \frac{9}{4}$
   (g) $x^2 + 2x + y^2 + 4y = 11$ (h) $x^2 - x + y^2 - 2y = 1$
   (i) $x^2 + 5x + y^2 + 3y = \frac{1}{2}$ (j) $x^2 - 10x + y^2 - 12y = 3$
   (k) $x^2 - 5x + y^2 = 0$ (l) $x^2 + 11x + y^2 - 20y = 15$

4. Draw graphs of the following equations.
   (a) $x^2 + y^2 = 49$ (b) $x^2 + y^2 = \frac{16}{9}$
   (c) $(x - 3)^2 + (y - 4)^2 = 49$ (d) $(x + 2)^2 + (y - 5)^2 = 49$
   (e) $(x - 4)^2 + (y - 3)^2 = \frac{16}{9}$ (f) $(x - \frac{1}{3})^2 + (y + \frac{1}{2})^2 = \frac{16}{9}$
   (g) $x^2 - 4x + y^2 + 6y = 36$ (h) $x^2 + 2x + y^2 + 3y = 3$
   (i) $x^2 - 8x + y^2 - 14y = 16$ (j) $x^2 - 7x + y^2 + 9y = -\frac{9}{4}$
   (k) $x^2 - 11x + y^2 = 0$ (l) $x^2 + 13x + y^2 - 12y = 15$

5. What is the graph of $x^2 + y^2 = 0$?
6. What is the graph of $x^2 + y^2 = -4$?
7. Must the graph of an equation of the form

$$x^2 + Bx + y^2 + Dy = E$$

always be a circle? Explain and justify your answer.

6. Graphs and Systems of Equations

In Chapter 2 we dealt with systems of equations such as

$$2x + 3y = 7$$
$$3x - y = 8$$

and determined algebraic methods of solving such systems. A graphic analysis of such systems gives some insight into their solutions.

Example 1

The graphs of the equations

$$2x + 3y = 7 \quad \text{and} \quad 3x - y = 8$$

both are lines and have been drawn together on the same coordinate system in Figure 6-23. From the graph we see that these lines meet at the point (2, 1). Since (2, 1) is the only point that is on both lines, it is the only solution to the system. $\square$

Figure 6-23

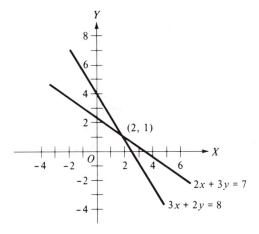

In general, the graphs of two equations of the form

$$ax + by = p \quad \text{and} \quad cx + dy = q$$

are lines. If the two lines do not intersect, that is, if they are parallel, then the system has no solution. If they meet at one point, then the system has one solution. If the two lines meet at more than one point, then they are the same line and every solution of one is a solution of the other. This justifies the assertion made in Chapter 2 that a system such as the above has one solution, no solution, or infinitely many solutions.

Thus far in this text we have limited ourselves to systems of linear equations, that is, equations in which no variable appears in a denominator, no variables are multiplied together, and the only powers of variables appearing are the first powers. Certainly we can imagine more complicated systems of equations.

Example 2

Solve the system

$$x^2 + y^2 = 16$$
$$2x - y = 1$$

Figure 6-24

The graph of the first equation is a circle and the graph of the second is a line. Thus either they do not meet and the system has no solution, or else they meet at one point or two points. (See Figure 6-24.)

Geometry and our knowledge of graphs assures us that there are 0, 1, or 2 solutions and no more. If we draw the correct graphs as in Figure 6-25, we see that they meet at two points. If we have very carefully drawn graphs (perhaps done by a computer) we could estimate these solutions from the graph. In general, however, we use algebraic techniques to find them. In this case we can solve the second equation for y in terms of x and obtain

$$y = 2x - 1$$

Figure 6-25

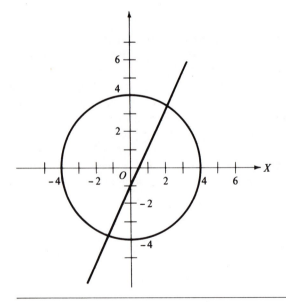

Substituting this in the first equation for y we have

$$x^2 + (2x - 1)^2 = 16$$
$$x^2 + 4x^2 - 4x + 1 = 16$$
$$5x^2 - 4x - 15 = 0$$

Using the quadratic formula,

$$x = \frac{4 \pm \sqrt{16 + 300}}{10}$$

$$= \frac{4 \pm \sqrt{316}}{10}$$

$$= \frac{4 \pm 2\sqrt{79}}{10}$$

$$= \frac{2 \pm \sqrt{79}}{5}$$

For

$$x = \frac{2 + \sqrt{79}}{5}$$

$$y = 2\left(\frac{2 + \sqrt{79}}{5}\right) - 1 = \frac{-1 + 2\sqrt{79}}{5}$$

For

$$x = \frac{2 - \sqrt{79}}{5}$$

$$y = 2\left(\frac{2 - \sqrt{79}}{5}\right) - 1 = \frac{-1 - 2\sqrt{79}}{5}$$

Thus the solutions are

$$\left(\frac{2 + \sqrt{79}}{5}, \frac{-1 + 2\sqrt{79}}{5}\right) \quad \text{and} \quad \left(\frac{2 - \sqrt{79}}{5}, \frac{-1 - 2\sqrt{79}}{5}\right) \quad \square$$

In solving more complicated systems of equations, you can expect to do more complicated algebraic manipulations. If, however, you have good graphs of the systems, you are likely to know what to expect as to the number of solutions and their locations.

Example 3

Solve the system

$$x^2 + (y - 2)^2 = 25$$
$$x^2 + y^2 = 9$$

Both graphs are circles, the radii are different, and they are disjoint or meet at one or two points. Subtracting yields

$$(y - 2)^2 - y^2 = 16$$
$$y^2 - 4y + 4 - y^2 = 16$$
$$-4y + 4 = 16$$
$$-4y = 12$$
$$y = -3$$

Thus any solutions must be on the line $y = -3$. If we substitute into the second equation we have

$$x^2 + 9 = 9$$
$$x^2 = 0$$
$$x = 0$$

Thus $(0, -3)$ is the only solution of the system. The graphs are given in Figure 6-26. □

More complicated systems may have no solutions.

Figure 6-26

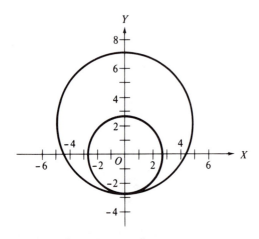

Example 4

Consider the system

$$3x + y = 20$$
$$x^2 + y^2 = 9$$

The two graphs are a line and a circle and, therefore, there are 0, 1, or 2 solutions. Solving the first equation for y we have

$$y = -3x + 20$$

Substituting in the second equation yields

$$x^2 + (-3x + 20)^2 = 9$$
$$x^2 + 9x^2 - 120x + 400 = 9$$
$$10x^2 - 120x + 391 = 0$$

Using the quadratic formula, we have

$$x = \frac{120 \pm \sqrt{(120)^2 - 4(10)(391)}}{20}$$

$$= \frac{120 \pm \sqrt{14{,}400 - 15{,}640}}{20}$$

$$= \frac{120 \pm \sqrt{-1240}}{20}$$

Since $\sqrt{-1240}$ does not exist, the system has no solutions. □

EXERCISES

Draw the graphs of the following systems of equations and find the solutions if solutions exist.

1. $5x + 3y = 8$
   $2x - 4y = 5$
2. $8x + 9y = 11$
   $2x + 3y = 8$
3. $2x + y = 2$
   $x^2 + y^2 = 16$
4. $x + 2y = 4$
   $x^2 + y^2 = 25$
5. $\quad\quad\quad 2x + 3y = 6$
   $(x - 2)^2 + (y - 1)^2 = 16$
6. $\quad\quad\quad x^2 + y^2 = 16$
   $(x - 1)^2 + y^2 = 16$
7. $(x - 2)^2 + (y - 3)^2 = 25$
   $(x + 1)^2 + (y + 2)^2 = 49$
8. $(x + 2)^2 + (y - 4)^2 = 25$
   $(x - 6)^2 + (y + 7)^2 = 9$
9. $3x + y = 2$
   $x^2 - y = 0$
10. $x^2 + y^2 = 25$
    $x^2 - y = 0$
11. $y = x^2 + 4x + 4$
    $y = x^2 - 6x + 9$
12. $y = 2x^2 + 4x + 2$
    $y = x^2 - 2x + 9$

7. Functions

The words "function" and "mapping" are used interchangeably in modern mathematics and, although we will use the word "function" most of the time in this text, the word "mapping" may give a better intuitive feeling for the concept. Suppose we have two sets of numbers X and Y, and a rule or process by which each element of X is associated with exactly one element of Y. This rule or process can be said to "map" X into Y. If we denote the process by a letter, such as f, a diagram such as Figure 6-27 gives us an intuitive grasp of the notion.

If X and Y are both R, the set of all real numbers, we could use number lines and draw a diagram as in Figure 6-28.

Figure 6–27

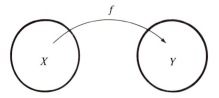

Figure 6–28

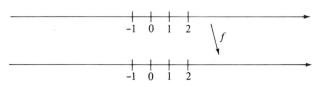

Example 1

Let f be the process that associates each real number with its square. (See Figure 6-29.) ☐

Figure 6–29

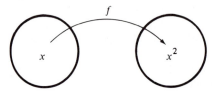

Example 2

Let g be the process that associates each number with its cube root. (See Figure 6-30.) ☐

Figure 6–30

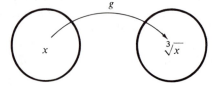

Example 3

Let h be the process that associates each number with its double. (See Figure 6-31.) □

Figure 6-31

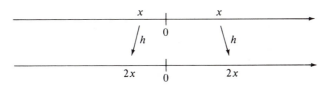

If f is a function and a is a number, we let $f(a)$ denote the unique number that f associates with a or that f "maps a to." We can use algebraic expressions to describe many mappings and the usual method is to use a variable, such as x, and an algebraic expression involving the variable to indicate the number that each value of x is associated with.

Thus for the function in Example 1 we could write

$$f(x) = x^2$$

For Example 2 we could write

$$g(x) = \sqrt[3]{x}$$

and for Example 3 we could write

$$h(x) = 2x$$

To determine the number that such a function, say k, associates with a number, we just substitute that number for x in the algebraic expression.

Example 4

Let

$$k(x) = 2x^2 - 3$$

Thus

$$k(0) = 0 - 3 = -3$$
$$k(1) = 2 \cdot 1 - 3 = -1$$
$$k(5) = 2 \cdot 25 - 3 = 47 \quad □$$

Example 5

Let

$$p(x) = \frac{x^2 - 7x}{x^2 + 1}$$

Then

$$p(0) = 0$$
$$p(1) = -3$$
$$p(4) = -\frac{12}{17} \quad \square$$

Example 6

Let

$$q(x) = \sqrt{x} \qquad \text{for all } x \geq 0$$

Then

$$q(0) = 0$$
$$q(4) = 2$$
$$q(19) = \sqrt{19} \quad \square$$

The set of all numbers that a function f "acts upon" or "maps to some number" is called the *domain* of f and the set of all numbers that f "maps something to" is the *range* of f. It is customary to assume, unless otherwise stated, that when a function is given by means of an algebraic expression involving a variable, say x, the domain of the function is the set of all numbers which can meaningfully be substituted for x.

Example 7

Let

$$s(x) = \sqrt{x - 3}$$

The domain of s is all real numbers greater than or equal to 3. $\square$

Example 8

Let

$$w(x) = \frac{4}{x^2 - 1}$$

The domain of w is all real numbers except 1 and -1. $\square$

Example 9

Let

$$z(x) = \frac{x^2 - 2x + 5}{x^2 + 4}$$

The domain of z is all real numbers. □

If we wish to use algebraic expressions to describe functions that have a certain set as the domain, all we need do is specify the domain in our description.

Example 10

Let

$$v(x) = x^2 + 3x - 2 \qquad \text{for } x \geq 5$$

In this case we have, by our description, limited the domain to numbers greater than or equal to 5. □

Example 11

Let

$$u(x) = x^2 + 1 \qquad \text{for } x \text{ an integer}$$

In this case, the domain is the set of integers. □

The range of a function described by an algebraic expression is more difficult to determine. We need to treat each function individually but some of the simpler ones are rather easy to analyze.

Example 12

Let

$$m(x) = x^2 + 5$$

Since $x^2 \geq 0$ for all x, $x^2 + 5 \geq 5$ for all x. Furthermore, $x^2 + 5$ will assume all possible values greater than or equal to 5 and therefore the range of m is the set of numbers greater than or equal to 5. □

Example 13

Let

$$n(x) = \sqrt[3]{x}$$

Since each real number has exactly one cube, each real number is the cube root of some number and the range of $n(x)$ is the set of all real numbers. □

The intuitive notion of a function can be made precise by the following formal set-theoretic definition:

A function is a set of ordered pairs of numbers in which no two distinct ordered pairs have the same first element.

This meets our intuitive notion if we consider each first number in an ordered pair to be associated with the second number in the pair. The domain is the set of all numbers which appear as first elements and the range is the set of all numbers which appear as second elements. We will not use this definition to any great extent but it does put the concept of a function on a more firm foundation.

The calculus and, more generally, real analysis are primarily concerned with the study of functions and one of the more useful tools for studying a function is its graph. The graph of a function, like the graph of an equation, is a picture of a set of ordered pairs. Since a function is a set of ordered pairs, the graph of a function is, in fact, a picture of the function. In order to draw the graph of a function, we draw the graph of the equation

$$y = f(x)$$

Example 14

To draw the graph of

$$f(x) = x^2 + 2x + 3$$

we draw the graph of

$$y = x^2 + 2x + 3$$

as illustrated in Figure 6-32. ☐

Figure 6-32

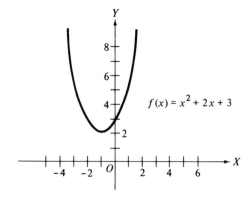

Example 15

To draw the graph of

$$g(x) = \frac{1}{x}$$

we draw the graph of

$$y = \frac{1}{x}$$

as in Figure 6-33. Notice that the domain of g is all nonzero real numbers. ☐

Figure 6-33

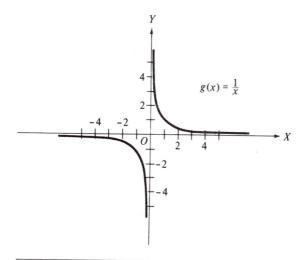

Our techniques for graphing functions are, at this point, limited to our techniques for graphing equations. If a function can be described using an algebraic expression which is related to some of the special forms we have previously discussed, then we can use our knowledge to draw the graph as we did in Examples 14 and 15. Otherwise we must use the "plot and sketch" technique. A function is said to be linear if its graph is a line.

Example 16

If

$$p(x) = 2x - 1$$

the graph of p is a line with slope 2 and Y-intercept -1. (See Figure 6-34.) ☐

Figure 6-34

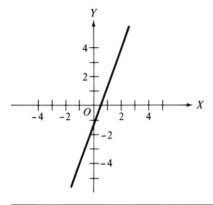

Example 17

If

$$h(x) = \sqrt{16 - x^2}$$

the graph of h is the graph of the equation

$$y = \sqrt{16 - x^2}$$

Squaring both sides yields

$$y^2 = 16 - x^2$$

Figure 6-35

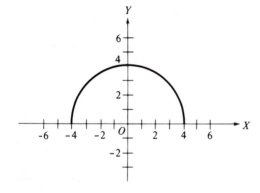

which is equivalent to

$$x^2 + y^2 = 16$$

This is the equation of a circle with center at $(0, 0)$ and radius 4. Since the original equation is

$$y = \sqrt{16 - x^2}$$

the only admissible values for y are nonnegative and the graph consists of the upper semicircle. The graph is given in Figure 6-35. □

Example 18

If

$$k(x) = \frac{4}{x^2 + 1}$$

k has the same shape on both sides of the Y-axis because $k(a) = k(-a)$ for each real number a. Since $x^2 \geq 0$ for all x, 1 is the smallest value of the denominator. Therefore, 4 is the largest value of $k(x)$. As we choose x larger, $k(x)$ gets smaller. The graph is given in Figure 6-36. □

Figure 6-36

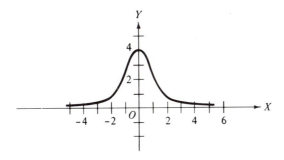

With the limited tools available, most graphs we draw must be fairly rough approximations but there are a few general procedures we can follow:

(i) Examine the function f to see if its equation $y = f(x)$ has some familiar form. If so, use your previous knowledge to draw it.

(ii) Otherwise try to determine sufficiently many points to give an outline of the general picture and then sketch from the outline. Among the more useful points are those where $f(x) = 0$ if such points exist. Such points are called the *zeros* of the function. It is also helpful to determine just where $f(x) > 0$ and where $f(x) < 0$.

Example 19

$$f(x) = x^3 - 5x^2 + 6x$$

Factoring yields

$$f(x) = x(x - 3)(x - 2)$$

Thus

$$f(x) = 0 \qquad \text{if } x = 0, 2, \text{ or } 3$$

and the graph crosses the X-axis at $(0, 0)$, $(2, 0)$, and $(3, 0)$.

Also

$$f(x) > 0 \qquad \text{if } x > 3 \quad \text{or} \quad 0 < x < 2$$

and

$$f(x) < 0 \qquad \text{if } x < 0 \quad \text{or} \quad 2 < x < 3$$

Using this information we can sketch a graph as in Figure 6-37. □

Figure 6–37

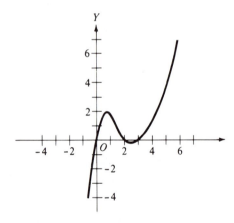

EXERCISES

1. Determine the domain and range of each of the following functions.

(a) $f(x) = \sqrt{25 - x^2}$

(b) $g(x) = \dfrac{4}{(x - 1)}$

(c) $h(x) = \dfrac{18}{(x^2 + 2)}$

(d) $k(x) = x^3 + 3x^2 - 2x + 1$

(e) $p(x) = \dfrac{x^3 + 5x^2 + 2x + 7}{x^2 - 3x - 4}$ (f) $q(x) = \dfrac{\sqrt{x^2 - 5}}{x^2 - 16}$

2. Determine the domain and range of each of the following functions.
   (a) $f(x) = \sqrt{49 - x^2}$ (b) $g(x) = 5/(x - 2)$
   (c) $h(x) = 12/(x^2 + 3)$ (d) $k(x) = 2x^3 - 7x^2 - 5x + 4$

   (e) $p(x) = \dfrac{x^3 + 6x^2 - 7x + 11}{x^2 - 7x + 12}$ (f) $q(x) = \dfrac{\sqrt{x^2 - 3}}{x^2 - 25}$

3. Draw the graphs of the following functions.
   (a) $f(x) = x^2 - 5x - 6$ (b) $g(x) = \sqrt{9 - x^2}$
   (c) $h(x) = \sqrt{9 + x^2}$ (d) $k(x) = x^3 + 7x^2 + 12x$
   (e) $p(x) = -\sqrt{16 - 4x^2}$ (f) $q(x) = (x + 4)/\sqrt{3}$
   (g) $r(x) = 6/(x^2 + 3)$ (h) $t(x) = \sqrt{5 - x^2 - 4x}$
   (i) $a(x) = |x|$ (j) $s(x) = \sqrt{x}$

4. Draw the graphs of the following functions.
   (a) $f(x) = x^2 - 7x + 10$ (b) $g(x) = \sqrt{16 - x^2}$
   (c) $h(x) = \sqrt{16 + x^2}$ (d) $k(x) = x^3 - 5x^2 + 6x$
   (e) $p(x) = -\sqrt{36 - 4x^2}$ (f) $q(x) = (x - 7)/\sqrt{2}$
   (g) $r(x) = 8/(x^2 + 4)$ (h) $t(x) = \sqrt{8 - x^2 - 2x}$
   (i) $a(x) = |x + 1|$ (j) $s(x) = \sqrt[3]{x}$

5. Draw the graphs of the following functions.
   (a) $f(x) = 1/x$ (b) $g(x) = \sqrt{x^2 - 6x}$
   (c) $h(x) = x/(x^2 + 1)$ (d) $p(x) = x^4 - 13x^2 + 36$
   (e) $k(x) = x^3 + 3x^2 - 4x - 12$ *(f) $q(x) = x^4 - 2x^3 - 9x^2 + 18x$

6. Draw the graphs of the following functions.
   (a) $f(x) = 3/x$ (b) $g(x) = \sqrt{x^2 - 10x}$
   (c) $h(x) = x/(x^2 + 4)$ (d) $p(x) = x^4 - 29x^2 + 100$
   (e) $k(x) = x^3 - 2x^2 - x + 2$ (f) $q(x) = x^4 - 5x^3 - 9x^2 + 45x$

7. Draw the graphs of the following functions. What can be said about their shapes?
   (a) $f(x) = x^3$ (b) $g(x) = (x - 1)^3$
   (c) $h(x) = (x + 2)^3$ (d) $t(x) = x^3 + 2$
   (e) $k(x) = (x - 1)^3 + 4$

8. Draw the graphs of the following functions. What can be said about their shapes?
   (a) $f(x) = x^4$ (b) $g(x) = (x + 1)^4$
   (c) $h(x) = (x - 2)^2$ (d) $t(x) = x^4 - 1$
   (e) $k(x) = (x + 1)^4 + 3$

9. Draw the graphs of the following functions. What can be said about their shapes?
   (a) $p(x) = 1/x^2$ (b) $q(x) = 1/(x + 1)^2$
   (c) $w(x) = 1/(x - 2)^2$ (d) $s(x) = (1/x^2) + 4$
   (e) $v(x) = (1/(x + 1)^2) + 2$

10. Draw the graphs of the following functions. What can be said about their shapes?
    (a) $p(x) = 1/x^3$ (b) $q(x) = 1/(x - 1)^3$
    (c) $w(x) = 1/(x + 3)^3$ (d) $v(x) = (1/(x - 1)^3) + 3$

*11. (a) Let $f(x)$ be a function and a a number. How are the graphs of $f(x)$ and
$g(x) = f(x - a)$ related?

(b) If b is a number, how are the graphs of $f(x)$ and $h(x) = f(x) + b$
related?

*12. If $p(x)$ is a function and a and b are numbers, how are the graphs of $p(x)$
and $q(x) = p(x - a) + b$ related?

13. Draw the graphs of the following functions.

(a) $f(x) = x$ (b) $f(x) = x^2$ (c) $g(x) = x^3$ (d) $h(x) = x^4$
(e) $k(x) = x^5$ (f) $m(x) = x^6$ (g) $n(x) = x^7$ (h) $p(x) = x^8$

14. Using Exercise 13 above, what can be said about the graph of $F(x) = x^n$ for
x a positive integer?

15. On the same coordinate system, draw the graphs of $f(x) = x^2$, $g(x) = 2x$, and
$h(x) = x^2 + 2x$. How are these graphs related?

16. On the same coordinate system, draw the graphs of $f(x) = 2x^2$, $g(x) = 3x - 1$,
and $h(x) = 2x^2 + 3x - 1$. How are these graphs related?

17. On the same coordinate system, draw the graphs of $f(x) = 5x^2$, $g(x) = 3x^2$,
and $h(x) = 2x^2$. How are these graphs related?

*18. (a) In general, how are the graphs of $f(x)$, $g(x)$, and $p(x) = f(x) + g(x)$ related?

(b) In general, how are the graphs of $f(x)$, $g(x)$, and $q(x) = f(x) - g(x)$ related?

(c) Suppose you are given the graphs of $f(x)$ and $g(x)$ on the same coordinate
system. Can you describe a method for drawing the graph of
$p(x) = f(x) + g(x)$? The graph of $q(x) = f(x) - g(x)$?

8. Applications of Functions

Our intuitive notion of a function is as an assignment process and the
everyday world abounds with such assignments. For example, the assignment of prices to goods in a store can be viewed as a function in that
every object is assigned a price. Frequently we are concerned with quantities
that depend on one another. For example, the amount of profit a business
concern earns depends on the amount of sales. We could, in fact, view
profit P as a function of sales S and write $P = f(S)$ or perhaps just denote
this function by $P(S)$.

Example 1

If a company makes a profit of 7 percent of sales, then the profit
function is given by

$P(S) = .07S$ $\square$

Example 2

If a manufacturing company produces a certain type of casting at a
cost of $17.50 per unit in material and labor and x is the number of

units produced, then the direct manufacturing costs are a function of the number of units produced and can be given by

$$C(x) = 17.50x \quad \square$$

Example 3

If the population of a city doubles every 25 years, and the population now is 100,000, then the population t years from now is given by

$$p(t) = 100,000 \ 2^{t/25} \quad \square$$

In each of the above examples, we expressed one quantity as a function of another. We can use graphs to analyze the behavior of such functions.

Example 4

Suppose the profit of a business can be given by

$$p(x) = .3x - .00001x^2 - 500$$

where x is the sales per month in dollars. If we draw the graph as in Figure 6-38, we see that the profits increase as sales increase to $15,000 and decrease after $15,000. From this it looks as if maximum profits occur if sales are $15,000 per month. $\square$

At this point, a philosophical remark is in order concerning the application of mathematics to real-life problems. Whenever we describe the relationship between two quantities by means of an algebraic or other mathematical equation, we are in fact *approximating* the relationship. In a great many

Figure 6-38

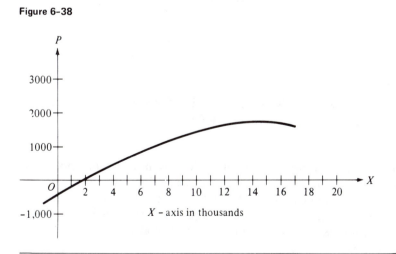

X - axis in thousands

cases the only values of x that we would consider in a real problem are integral or whole number values. In Example 4, we would not be concerned with sales of $4397.46 but would probably list sales to the nearest thousand. If we did so, the graph would look somewhat like Figure 6-39.

It is simpler, however, to work with functions that are defined for all real numbers or at least for a suitable interval of real numbers. The reader should understand that answers to real problems must be reasonable. For example, it is meaningless to say that an auto dealership will maximize its profits by selling 42.8 cars per month. What we really mean is that sales of 42 or 43 cars per month will maximize profits.

All of the functions and equations used to describe real situations should be viewed as suitable but imperfect approximations and the answers obtained by using these must also be approximate.

Figure 6-39

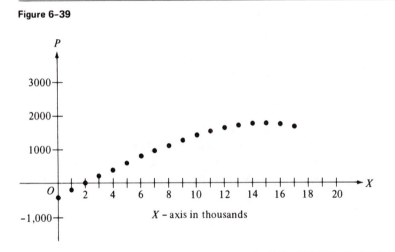

X – axis in thousands

Example 5

The city income tax in Bowling Green is 1.5 percent of gross earnings. We can approximate the tax function $T(x)$ by

$$T(x) = .015x$$

where x is the gross earnings in dollars. This is, however, an approximation since the only values x can assume are finite decimals with two places. (You cannot earn $6434.7892 and neither can you earn π.) Instead of considering only these values, however, we will view T as a function with the nonnegative numbers for its domain; its graph is given by Figure 6-40. □

Figure 6-40

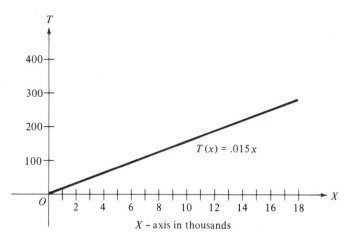

The simplest types of functions to analyze are linear functions. If we know that one quantity is a linear function of another, we need only two values to determine the function. If a quantity or variable p is a linear function of a quantity or variable q, then we know p *varies linearly* with q in the sense discussed in Section 4.

Example 6

Suppose we know that in a certain business, profits vary linearly with sales. Then, for some numbers m and b,

$$P(S) = mS + b$$

that is, profit is a linear function of sales. To obtain m and b we need only to know two sets of corresponding values. Suppose profits are $1000 on sales of $12,000 and are $1500 on sales of $16,000. The graph of P is a line through the points $(12{,}000, 1000)$ and $(16{,}000, 1500)$. From previous work we know that the slope m is given by

$$\frac{1500 - 1000}{16{,}000 - 12{,}000} = \frac{500}{4000} = \frac{5}{40} = \frac{1}{8}$$

Thus

$$P(S) = \tfrac{1}{8}S + b$$

and we have yet to find b. Again using the first given points we have

$$1000 = \tfrac{1}{8}(12{,}000) + b$$
$$= 1500 + b$$
$$b = -500$$

Thus

$$P(S) = \tfrac{1}{8}S - 500$$

The graph is given in Figure 6-41. □

Figure 6-41

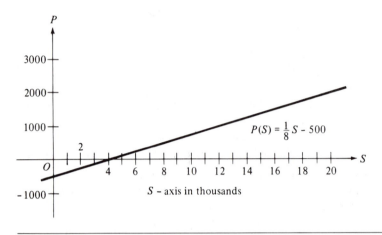

In general, if we have a linear function

$$f(x) = mx + b$$

we can find m by using two points (or sets of values) and then use one of these points to find b.

Example 7

Suppose, in a certain manufacturing concern, costs are known to be a linear function of the number of units produced. If $C = \$4200$ when 700 units are produced and $C = \$4700$ when 800 units are produced, then

$$C(x) = mx + b$$

for some numbers m and b. It follows that

$$4200 = 700m + b$$
$$4700 = 800m + b$$

and

$$500 = 100m$$

Therefore

$m = 5$

Since

$m = 5$
$4200 = 5(700) + b$
$4200 = 3500 + b$
$\quad b = 700$

Hence

$C(x) = 5x + 700 \;\square$

Since linear functions are the easiest types of functions to deal with, it is common in applications to try to approximate the relationships between quantities by means of a linear function.

A function is *piecewise linear* if its graph is made up of line segments. Such functions are illustrated in Figure 6-42. An example of a piecewise linear function from the real world is the tax function for the so-called progressive income tax. This is partially illustrated in Figure 6-43.

If we have several sets of values for two quantities, the functional relationship between them can be approximated by a piecewise linear function just by plotting the points and connecting them with line segments.

Figure 6-42

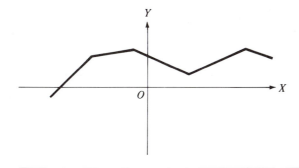

Example 8

Suppose the costs for a manufacturing concern are a function of the number of units produced and the following production and cost data are known.

Number of units	100	200	300	400
Costs	1000	1600	2000	2200

Figure 6-43

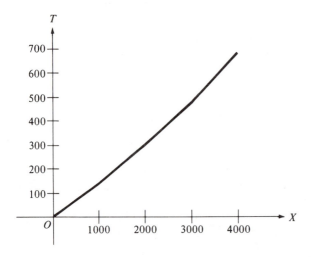

If we plot these points on a coordinate system and connect them with line segments, we have approximated the cost function as a piecewise linear function. Note also, however, that the function

$$C(x) = 9x - .01x^2 + 200$$

also passes through these points. (See Figure 6-44.) Thus the actual cost function is approximated by both the piecewise linear function and the polynomial function. In order to describe the piecewise linear function we need to give quite a complicated description.

Figure 6-44

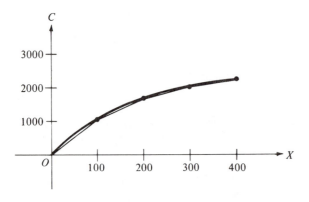

$$C(x) = 6x + 400 \qquad \text{for } 100 \leq x \leq 200$$
$$= 4x + 800 \qquad \text{for } 200 \leq x \leq 300$$
$$= 2x + 1400 \qquad \text{for } 300 \leq x \leq 400$$

If we use the polynomial approximation we have simply

$$C(x) = 9x - .01x^2 + 200 \qquad \text{for } 100 \leq x \leq 400 \quad \square$$

Remember, however, that functions such as $C(x)$ above are always approximations of the real relationship and a piecewise linear function will provide as good an approximation providing we have enough points. After all, all real-world functional relationships consist of only finitely many points and a piecewise linear approximation connecting these points is the best we could hope for. It would be clumsy to plot all the values and, in fact, in most cases we *do not know* all the values. We are constructing a function from a few values in the hope that it will help us to study all the values.

Not all functions have graphs that are nicely connected sets of points as those in the previous examples.

Example 9

The cost of mailing is a function of the weight of the object to be mailed (providing the class of service is fixed). We could write

$$C = P(w)$$

where w denotes the weight and P is the postage function. The graph is given in Figure 6-45 for first class postage; the weight is given in ounces

Figure 6-45

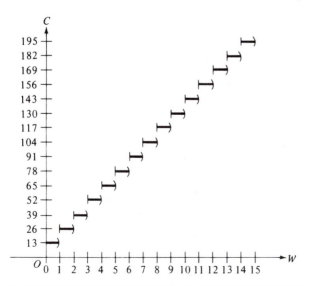

and the cost in cents. Notice that the graph is not a line or a smooth curve but is, in fact, a sequence of steps. ☐

Frequently it is useful to approximate a function by a linear function because of the simplicity of dealing with the linear function.

Example 10

The gasoline mileage obtained in expressway driving will depend on the speed driven and, for a particular individual and his car, can be viewed as a function of the speed. Suppose you make five trips, over the same road, with your car and obtain the following results:

Trip number	1	2	3	4	5
S = avg. speed	50	55	60	65	70
M = avg. mpg	15.0	14.2	13.5	12.7	12.0

The graph showing these five points is given in Figure 6-46. It appears that they almost fall on the same line. In fact, the line

$$M - 15 = -.15(S - 50)$$

Figure 6-46

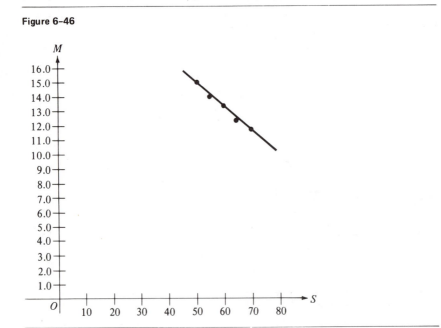

provides a good approximation of the graph for values of S between 50 and 70. We might even say that the mileage function is given, approximately, by

$$M = M(s) = -.15S + 22.5$$

for values of S between 50 and 70. In this case, we have approximated the actual mileage function by a nice simple linear function. $\square$

EXERCISES

1. Determine costs as a function of units produced if you know that costs are a linear function of the number of units produced, that the cost of producing 100 units is $1050, and the cost of producing 400 units is $3300. What is the cost of producing 275 units?

2. Determine profits as a function of sales if profits are a linear function of sales, the profit on sales of $30,000 is $300, and the profit on sales of $50,000 is $900. What is the profit on sales of $72,000?

3. It might be expected that the number of enlistments in the armed forces is, in some way, a function of the unemployment rate in a locality. (As jobs become harder to find, more people want the security of the armed forces.) Suppose experience has shown that this is a linear relationship, that is, the number of enlistments is a linear function of the unemployment rate in a particular city. If 160 men join during a year when the average unemployment rate is 5 percent and 148 men join during a year when the average unemployment rate is 4 percent, determine the number of enlistments as a function of the unemployment rate. How many enlistments would you expect with a rate of 6.5 percent?

4. Fixed costs of a business are costs that do not change with the number of goods produced; for example, taxes on the business property and (possibly) executive salaries. If costs are a linear function of the number of units produced, what point on the graph gives the fixed costs?

5. If profit is a function of sales, what would experience lead you to expect about the P-intercept of the function, that is, what kind of number would you expect $P(0)$ to be?

6. The revenue R of a business is the total income and the profit P is given by $P = R - C$ where C represents costs. If x is the number of units sold at a price of t dollars (t may also vary), then $R = tx$.
   (a) Under what condition is R a linear function of x?
   (b) If R and C are linear functions of x, is P a linear function of x? What can you say about the slope of the graph of P (providing the slope exists)?

7. Use the following data to draw a piecewise linear graph approximating each of the given functions.

(a)

x	1	2	3	4	5	6
$f(x)$	4	7	5	2	6	5

(b)

x	−3	0	1	2	3	4
g(x)	2	4	−2	5	4	2

(c)

t	0	1	2	3	4	5
p(t)	0	1	4	9	16	25

(d)

x	0	−1	1	−2	2
q(x)	1	0	2	−7	9

(e)

t	0	−1	1	−2	$\frac{3}{2}$	3
h(t)	2	6	0	12	$-\frac{1}{4}$	2

8. Find a linear function that approximates the function F if the following table gives points on the graph of F.

x	0	1	2	3	4	6	10
F(x)	3.1	4.2	6	7.5	8.9	12.1	18

9. Find a linear function that approximates the function G if the following table gives points on the graph of G.

x	0	1	2	3	5	6	−3
G(x)	2	2.6	3.3	4	5.3	10	0

*10. Try to find a polynomial function that passes through the same points as the graphs of the functions p, q, and h given in Exercise 7.

*9. The Arithmetic of Functions

If f and g are functions and a is a number in the domains of both f and g, then $f(a)$ and $g(a)$ are numbers and we can perform the usual arithmetic operations with them, that is, we can compute

$$f(a) + g(a), \qquad f(a) - g(a), \qquad g(a) - f(a),$$

$$[f(a)][g(a)], \qquad \frac{f(a)}{g(a)} \quad [\text{if } g(a) \neq 0], \quad \text{and} \quad \frac{g(a)}{f(a)} \quad [\text{if } f(a) \neq 0].$$

Example 1

If $f(x) = x^2$ and $g(x) = x + 1$, then the number 4 is in the domain of both functions and

$$f(4) + g(4) = 16 + 5 = 21$$
$$f(4) - g(4) = 16 - 5 = 11$$
$$g(4) - f(4) = 5 - 16 = -11$$
$$[f(4)][g(4)] = (16)(5) = 80$$
$$\frac{f(4)}{g(4)} = \frac{16}{5}$$
$$\frac{g(4)}{f(4)} = \frac{5}{16} \quad \square$$

Example 2

If $p(x) = x - 7$ and $q(x) = \sqrt{x - 1}$, then $p(-2) + q(-2)$ has no meaning because $q(-2)$ does not exist; that is, -2 is not in the domain of q. Furthermore, $p(1)/q(1)$ has no meaning because $q(1) = 0$. $\square$

If f and g are functions, we can use these arithmetic processes to define new functions as follows:

(i) The *sum* of f and g is denoted by $f + g$ and defined by

$$(f + g)(x) = f(x) + g(x)$$

for each number x that is in both the domain of f and the domain of g.

(ii) The *difference* of f and g is denoted by $f - g$ and defined by

$$(f - g)(x) = f(x) - g(x)$$

for each number x that is in both domains.

(iii) The *product* of f and g is denoted by fg and defined by

$$(fg)(x) = f(x)g(x)$$

for each number x in both domains.

(iv) The *quotient* of f and g is denoted by f/g or $f \div g$ and defined by

$$\left(\frac{f}{g}\right)(x) = \frac{f(x)}{g(x)}$$

for each number x in both domains under the further restriction that $g(x) \neq 0$.

These operations are called, as expected, addition, subtraction, multiplication, and division of functions. They are not the same as the corresponding arithmetic operations for numbers but, because they are defined using these operations, they do inherit the associative and commutative properties for addition and multiplication and the distributive law. In fact, most of the basic arithmetic properties of the real number system extend to the arithmetic of functions. The constant function $z(x) = 0$ takes the place of the number 0 and the constant function $I(x) = 1$ behaves like the number 1.

If we have functions that are given by algebraic expressions, the simplest way to express the sum, product, difference, or quotient may be by indicating the sum, product, difference, or quotient of the original expressions and perhaps simplifying them. Care must be used in doing so to make sure that the new domain is specified since the resulting expression may have meaning for substitutions that were meaningless for one of the original functions.

Example 3

If $h(x) = 1/(x - 1)$ and $k(x) = \sqrt{x}$, then the domain of $h + k$ is the set of all nonnegative numbers except 1 and

$$(h + k)(x) = h(x) + k(x) = \frac{1}{x - 1} + \sqrt{x} = \frac{x\sqrt{x} - \sqrt{x} + 1}{x - 1}$$

The domain of the product is the same as that of the sum and

$$(hk)(x) = h(x)k(x) = \frac{1}{x - 1}\sqrt{x} = \frac{\sqrt{x}}{x - 1}$$

The difference also has this domain and

$$(h - k)(x) = \frac{1}{x - 1} - \sqrt{x} = \frac{1 - x\sqrt{x} + \sqrt{x}}{x - 1}$$

Also

$$\left(\frac{h}{k}\right)(x) = \frac{h(x)}{k(x)} = \frac{1}{x - 1}\sqrt{x} = \frac{1}{(x - 1)\sqrt{x}}$$

with the domain being the set of all positive numbers except 1. □

Example 4

If

$$f(x) = x^2 - 4 \quad \text{and} \quad g(x) = x - 2$$

then

$$(f + g)(x) = f(x) + g(x) = (x^2 - 4) + (x - 2) = x^2 + x - 6$$
$$(fg)(x) = (x^2 - 4)(x - 2) = x^3 - 2x^2 - 4x + 8$$
$$\left(\frac{f}{g}\right)(x) = \frac{f(x)}{g(x)} = \frac{x^2 - 4}{x - 2} = x + 2 \quad \text{for } x \neq 2$$

Although $x + 2$ has meaning for $x = 2$, $(x^2 - 4)/(x - 2)$ does not and the domain of f/g is the set of all numbers except 2. ☐

Example 5

If

$$p(x) = x^2 + \sqrt{x} \quad \text{and} \quad q(x) = \sqrt{x}$$

then

$$(p - q)(x) = (x^2 + \sqrt{x}) - \sqrt{x} = x^2$$

The domain of $p - q$, however, is the set of all nonnegative numbers because these are the only numbers for which p and q both have meaning. ☐

These arithmetic operations for functions can be used to simplify the graphing of functions. If we know the graph of f and the graph of g, we can draw the graph of $f + g$ just by adding Y-coordinates of points with the same X-coordinate. Similarly, the graphs of $f - g$, fg, and f/g can be drawn by subtracting, multiplying, or dividing these Y-coordinates.

The simplest cases are those in which one of the functions is a constant.

Figure 6–47

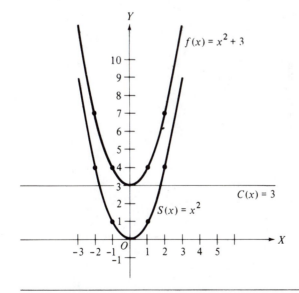

Example 6

To draw the graph of

$$f(x) = x^2 + 3$$

first notice that f is the sum of $s(x) = x^2$ and the constant function $c(x) = 3$.

In order to draw the graph of this function, we need only add 3 to each Y-coordinate on the graph of

$$s(x) = x^2$$

See Figure 6-47. □

Example 7

Draw the graph of

$$g(x) = x^3 - 1$$

We need only to subtract 1 from each Y-coordinate of the graph of

$$p(x) = x^3$$

See Figure 6-48. □

Figure 6–48

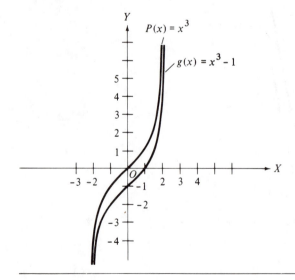

Figure 6–49

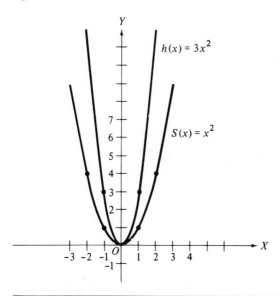

Example 8

The graph of

$$h(x) = 3x^2$$

can be drawn by multiplying each Y-coordinate of

$$s(x) = x^2$$

by 3. See Figure 6-49.. □

Example 9

The graph of

$$w(x) = x^3 + x^2$$

can be drawn by adding the Y-coordinates of the graphs of

$$p(x) = x^3 \quad \text{and} \quad q(x) = x^2$$

See Figure 6-50. □

Figure 6–50

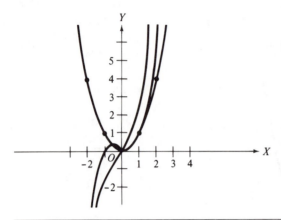

Example 10

The graph of

$$z(x) = \frac{x^3}{x^2 + 1}$$

can be drawn by dividing the Y-coordinates of

$$p(x) = x^3$$

by the Y-coordinates of

$$v(x) = x^2 + 1$$

See Figure 6-51. □

These methods are easiest to use when one of the functions is a constant or when adding or subtracting.
Remember:

(i) Adding or subtracting a constant function just shifts the curve upward or downward by the absolute value of the constant.
(ii) Multiplying or dividing by a constant function just distorts the curve vertically.

Composition is another very useful way to combine two functions to produce a new function. If f and g are functions, the *composite of f and g* or *f composed with g* is denoted by $f \circ g$ and defined by

$$(f \circ g)(x) = f(g(x))$$

Figure 6-51

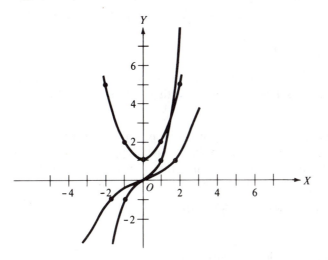

for each number x in the domain of g such that the number $g(x)$ is in the domain of f.

You have seen many functions which are, in fact, the result of composing functions.

Example 11

Let

$$s(x) = x^2 \quad \text{and} \quad g(x) = 2x - 1$$

Then

$$(s \circ g)(x) = s(g(x)) = s(2x - 1) = (2x - 1)^2$$

Since both domains are R, the domain of $s \circ g$ is also R. Notice that

$$(g \circ s)(x) = g(s(x)) = g(x^2) = 2x^2 - 1 \;\square$$

Example 12

Let

$$s(x) = x^2 \quad \text{and} \quad p(x) = \sqrt{x - 2}$$

Then

$$(s \circ p)(x) = s(p(x)) = s(\sqrt{x - 2}) = (\sqrt{x - 2})^2 = x - 2$$

Although the expression $x - 2$ has meaning for all substitutions, the domain of p is all numbers greater than or equal to 2. Since $p(x)$ is in the domain of $s(x)$ for all x, the domain of $s \circ p$ is just the domain of p. Notice that

$$(p \circ s)(x) = p(s(x)) = p(x^2) = \sqrt{x^2 - 2}$$

Since $\sqrt{x^2 - 2}$ has meaning only for values of x such that $|x| \geq \sqrt{2}$ the domain of $p \circ s$ is the set of all numbers with absolute value of at least $\sqrt{2}$. $\square$

From the examples, we see that, in general, $f \circ g \neq g \circ f$ and composition of functions does not satisfy the commutative law. Suppose now that f, g, and h are functions and a is a number such that $h(a)$ is in the domain of g and $g(h(a))$ is in the domain of f. Then

$$[(f \circ g) \circ h](a) = (f \circ g)(h(a)) = f(g(h(a)))$$

and

$$[(f \circ (g \circ h)](a) = f((g \circ h)(a)) = f(g(h(a)))$$

Thus for each number a where $f \circ (g \circ h)$ or $(f \circ g) \circ h$ has meaning,

$$[f \circ (g \circ h)](a) = [(f \circ g) \circ h](a)$$

Thus

$$[f \circ (g \circ h)] = [(f \circ g) \circ h]$$

and composition of functions satisfies the associative law.

Example 13

Let

$$f(x) = x^2, \quad g(x) = 2x + 1, \quad h(x) = \frac{1}{x - 1}$$

$$\begin{aligned}
[f \circ (g \circ h)](x) &= f((g \circ h)(x)) \\
&= [(g \circ h)(x)]^2 \\
&= [g(h(x))]^2 \\
&= [2(h(x)) + 1]^2 \\
&= \left[2\left(\frac{1}{x-1} \right) + 1 \right]^2 \\
&= \left[\frac{2}{x-1} + 1 \right]^2 \\
&= \left(\frac{x+1}{x-1} \right)^2
\end{aligned}$$

$$[(f \circ g) \circ h](x) = (f \circ g)(h(x))$$

$$= (f \circ g)\left(\frac{1}{x-1}\right)$$

$$= f\left[g\left(\frac{1}{x-1}\right)\right]$$

$$= f\left[2\left(\frac{1}{x-1}\right) + 1\right]$$

$$= f\left[\frac{x+1}{x-1}\right]$$

$$= \left(\frac{x+1}{x-1}\right)^2 \quad \square$$

In Exercise 7 of Section 4 you were asked to draw the graphs of $f(x) = x^3$, $g(x) = (x-1)^3$, and $h(x) = (x+2)^3$. All three of these graphs have the same shape, they are just located differently on the coordinate system. (See Figure 6-52.)

In general, the graphs of $f(x)$ and $f(x-a)$ have exactly the same shape. The function $f(x-a)$, after all, is just $(f \circ g)(x)$ where $g(x) = x - a$.

The result is to move the graph of f a units to the right if $a > 0$ and

Figure 6-52

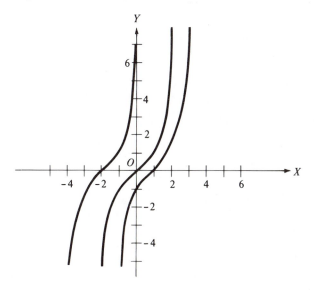

$|a|$ units to the left if $a < 0$. This can be seen by noticing that if b is any number, then

$$f(b) = f[(b + a) - a]$$

Thus $f(x - a)$ takes the same value at $b + a$ as $f(x)$ does at b and the graph of $f(x - a)$ is just the graph of $f(x)$ shifted $|a|$ units to the right or left, right if $a > 0$ and left if $a < 0$.

Example 14

Draw the graph of

$$f(x) = (x - 3)^3 + 1$$

The graph of $g(x) = x^3 + 1$ is drawn by adding 1 to each Y-coordinate of the graph of $y = x^3$. The graph of $f(x)$ is just the graph of $g(x)$ shifted 3 units to the right. See Figure 6-53. ☐

In summary:

(i) The graphs of $f + g$, $f - g$, fg, and f/g can be drawn using the graphs of f and g simply by adding, subtracting, multiplying, or dividing the two Y-coordinates associated with each number x. In practice, we compute enough of these values to enable us to sketch the graph. When one of the functions is a constant, the computation is quite simple.

(ii) The graph of $f(x - a)$ has the same shape as the graph of $f(x)$; it is just shifted right or left by $|a|$ units depending on whether a is positive or negative.

Figure 6-53

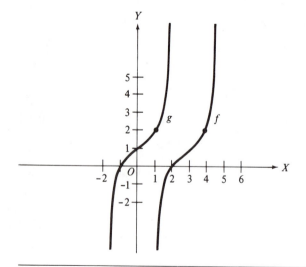

EXERCISES

1. Let $f(x) = x^2 + 1$, $g(x) = x^3$, $h(x) = \sqrt{x - 2}$, and $k(x) = 1/(x + 1)$. Determine simplified expressions for the following functions and specify the domain of each.
   (a) $f + g$ (b) fg (c) g/h (d) hk
   (e) fk (f) $g - f$ (g) g/k (h) k/f

2. Draw the graphs of the functions f, g, h, and k given in Exercise 1 and the graphs of those you were asked to express.

3. Let f, g, h, and k be as in Exercise 1. Determine simplified expressions for the following functions and specify the domain of each.
   (a) $f \circ g$ (b) $g \circ f$ (c) $f \circ h$ (d) $h \circ k$
   (e) $f \circ k$ (f) $g \circ h$ (g) $(h \circ f) \circ k$ (h) $k \circ (g \circ k)$

4. Use the method illustrated in Example 14 to draw the graphs of the following functions.
   (a) $w(x) = (x^2 - 3)^3$ (b) $v(x) = (x^3 + 2)^2$
   (c) $p(x) = \sqrt{2x - 7}$ (d) $q(x) = (3x + 2)^3$

5. Draw the graphs of the following functions.
   (a) $f(x) = x^2 - 5$ (b) $g(x) = (x - 4)^3 + 3$
   (c) $h(x) = (2x - 5)^4 - 1$ (d) $k(x) = \sqrt{(x - 7)^3 + 2}$
   (e) $p(x) = [\sqrt{x + 1} - 3]^4$ (f) $q(x) = 1/x^2$
   (g) $w(x) = 1/(x - 2)^2$ (h) $v(x) = x + 2/x$
   (i) $u(x) = x + 2/(x - 3)$ (j) $z(x) = x^2 + 1/x$

*10. Ellipses

In Section 5 we saw that an equation of the form

$$x^2 + y^2 = r^2$$

has as its graph a circle with center $(0, 0)$ and radius r.

Example 1

Consider the equation

$$x^2 + 4y^2 = 16$$

From Section 5, we know that the graph is not a circle, but when we plot some points and sketch the set, it looks like a "flattened" circle as in Figure 6-54. This curve is, in fact, an *ellipse* and the graph is a bit easier to draw if we first divide both sides of the equation by 16.

$$\frac{x^2}{16} + \frac{y^2}{4} = 1$$

Figure 6-54

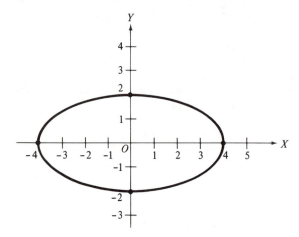

Then when $x = 0$,

$$\frac{y^2}{4} = 1 \quad \text{and} \quad y = \pm 2$$

When $y = 0$,

$$\frac{x^2}{16} = 1 \quad \text{and} \quad x = \pm 4$$

Thus the points $(0, 2)$, $(0, -2)$, $(4, 0)$, and $(-4, 0)$ are on the graph. From the equation

$$\frac{x^2}{16} + \frac{y^2}{4} = 1$$

we see that if (x, y) is on the graph, $|x|$ does not exceed 4 for otherwise the left side would be greater than 1. Similarly, $|y| \le 2$ or the left side will be greater than 1. Thus all points on the graph will be such that $-4 \le x \le 4$ and $-2 \le y \le 2$. The points $(0, 2)$, $(0, -2)$, $(4, 0)$, $(-4, 0)$ are called the *vertices* of the ellipse (each of these four points is a *vertex*) and since $(0, 0)$ is halfway between opposite vertices, $(0, 0)$ is called the *center* of this ellipse. ☐

In general, if $a > 0$ and $b > 0$, the graph of

$$\frac{x^2}{a^2} + \frac{y^2}{b^2} = 1$$

is an ellipse. When $x = 0$, $y = \pm b$ and when $y = 0$, $x = \pm a$. Thus the points $(0, b)$, $(0, -b)$, $(a, 0)$, and $(-a, 0)$ are on the graph. Furthermore, if (x, y) is on the graph, then

$$-a \le x \le a \quad \text{and} \quad -b \le y \le b$$

for otherwise the left side of the equation is greater than 1. The points $(0, b)$, $(0, -b)$, $(a, 0)$, and $(-a, 0)$ are the *vertices* of the ellipse and $(0, 0)$ is the *center*. We draw the graph by plotting the vertices, some other points if needed, and then draw the smooth curve containing them. Several graphs are given in Figure 6-55.

Figure 6-55

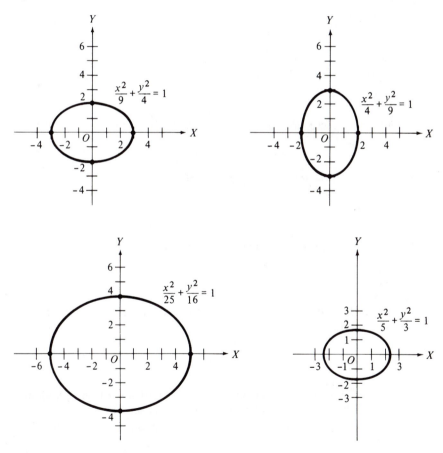

Example 2

Consider now the equation

$$\frac{(x-5)^2}{9} + \frac{(y-7)^2}{4} = 1$$

When $x = 5$, $y - 7 = \pm 2$ and $y = 9$ or $y = 5$. When $y = 7$, $x - 5 = \pm 3$ and $x = 8$ or $x = 2$. Furthermore, $|x - 5| \leq 3$ and $|y - 7| \leq 2$ or otherwise the left side is greater than 1. If we compute a few points and sketch the curve we see that it has the same shape as

$$\frac{x^2}{9} + \frac{y^2}{4} = 1$$

It is just located in a different place in the plane.

If we draw dotted vertical and horizontal lines through $(5, 7)$, we see that the graph of

$$\frac{(x-5)^2}{9} + \frac{(y-7)^2}{4} = 1$$

is just the graph of

$$\frac{x^2}{9} + \frac{y^2}{4} = 1$$

drawn using these dotted lines as the axes. Both graphs are given in Figure 6-56. □

Thus if we can draw graphs of equations of the form

$$\frac{x^2}{a^2} + \frac{y^2}{b^2} = 1$$

then we can draw graphs of equations of the form

$$\frac{(x-h)^2}{a^2} + \frac{(y-k)^2}{b^2} = 1$$

All we need do is draw a new dotted set of axes with the origin at (h, k) and draw the graph of

$$\frac{x^2}{a^2} + \frac{y^2}{b^2} = 1$$

on this "new" set of axes. The points $(h, k + b)$, $(h, k - b)$, $(h + a, k)$, and $(h - a, k)$ will be the vertices and (h, k) will be the center.

The completing the square technique can be used to transform an equation of the form

$$Ax^2 + Bx + Cy^2 + Dy = E \qquad \text{with } A > 0 \text{ and } C > 0$$

Figure 6-56

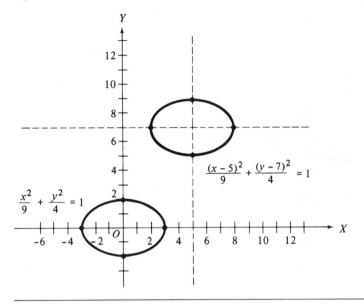

$$\frac{(x-5)^2}{9} + \frac{(y-7)^2}{4} = 1$$

$$\frac{x^2}{9} + \frac{y^2}{4} = 1$$

into one of the form

$$\frac{(x-h)^2}{a^2} + \frac{(y-k)^2}{b^2} = 1$$

if E is sufficiently large. (The number E must be large enough to assure that the right side of the resulting equation is positive.)

Example 3

$$4x^2 + 24x + 9y^2 - 18y = 4$$
$$4(x^2 + 6x) + 9(y^2 - 2y) = 4$$
$$4(x^2 + 6x + 9) + 9(y^2 - 2y + 1) = 4 + 36 + 9$$
$$4(x + 3)^2 + 9(y - 1)^2 = 49$$
$$\frac{4(x + 3)^2}{49} + \frac{9(y - 1)^2}{49} = 1$$
$$\frac{(x + 3)^2}{49/4} + \frac{(y - 1)^2}{49/9} = 1$$
$$\frac{(x + 3)^2}{(7/2)^2} + \frac{(y - 1)^2}{(7/3)^2} = 1$$

In this case $a = 7/2$, $b = 7/3$, $h = -3$, and $k = 1$. The center is at $(-3, 1)$ and the vertices are $(-13/2, 1)$, $(1/2, 1)$, $(-3, 10/3)$, and $(-3, -4/3)$. The graph is given in Figure 6-57. $\square$

Figure 6-57

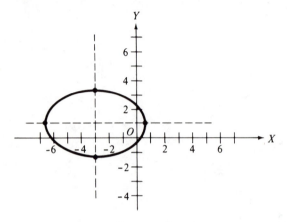

EXERCISES

1. Draw the graphs of the following equations.

(a) $\dfrac{x^2}{4} + \dfrac{y^2}{25} = 1$ (b) $\dfrac{x^2}{16} + \dfrac{y^2}{49} = 1$

(c) $\dfrac{x^2}{1} + \dfrac{y^2}{16} = 1$ (d) $\dfrac{x^2}{25} + \dfrac{y^2}{9} = 1$

(e) $\dfrac{x^2}{49} + \dfrac{y^2}{4} = 1$ (f) $\dfrac{x^2}{10} + \dfrac{y^2}{16} = 1$

(g) $\dfrac{x^2}{9} + \dfrac{y^2}{5} = 1$ (h) $\dfrac{x^2}{11} + \dfrac{y^2}{6} = 1$

2. Draw the graphs of the following equations.

(a) $\dfrac{x^2}{25} + \dfrac{y^2}{49} = 1$ (b) $\dfrac{x^2}{4} + \dfrac{y^2}{16} = 1$

(c) $\dfrac{x^2}{9} + \dfrac{y^2}{1} = 1$ (d) $\dfrac{x^2}{25} + \dfrac{y^2}{4} = 1$

(e) $\dfrac{x^2}{64} + \dfrac{y^2}{25} = 1$ (f) $\dfrac{x^2}{12} + \dfrac{y^2}{9} = 1$

(g) $\dfrac{x^2}{16} + \dfrac{y^2}{7} = 1$ (h) $\dfrac{x^2}{17} + \dfrac{y^2}{10} = 1$

3. Draw the graphs of the following equations.

(a) $\dfrac{(x-1)^2}{4} + \dfrac{(y-3)^2}{25} = 1$ (b) $\dfrac{(x+2)^2}{16} + \dfrac{(y-4)^2}{49} = 1$

(c) $\dfrac{(x+5)^2}{1} + \dfrac{(y-2)^2}{16} = 1$ (d) $\dfrac{(x-4)^2}{25} + \dfrac{(y+5)^2}{9} = 1$

(e) $\dfrac{(x+3)^2}{49} + \dfrac{(y-5)^2}{4} = 1$ (f) $\dfrac{(x+2)^2}{10} + \dfrac{(y-\frac{3}{2})^2}{16} = 1$

(g) $\dfrac{(x-4.1)^2}{9} + \dfrac{(y+1.3)^2}{5} = 1$ (h) $\dfrac{(x+1.3)^2}{11} + \dfrac{(y-4.7)^2}{6} = 1$

4. Draw the graphs of the following equations.

(a) $\dfrac{(x-2)^2}{25} + \dfrac{(y-3)^2}{49} = 1$ (b) $\dfrac{(x+1)^2}{4} + \dfrac{(y-5)^2}{16} = 1$

(c) $\dfrac{(x+5)^2}{9} + \dfrac{(y+7)^2}{1} = 1$ (d) $\dfrac{(x+4)^2}{25} + \dfrac{(y-5)^2}{9} = 1$

(e) $\dfrac{(x-4.5)^2}{64} + \dfrac{(y-2.3)^2}{25} = 1$ (f) $\dfrac{(x+7)^2}{12} + \dfrac{(y+8)^2}{9} = 1$

(g) $\dfrac{(x-\frac{1}{2})^2}{16} + \dfrac{(y+2)^2}{7} = 1$ (h) $\dfrac{(x-2.9)^2}{17} + \dfrac{(y-1.7)^2}{10} = 1$

5. Draw the graphs of the following equations.
(a) $x^2 - 6x + 4y^2 + 3y = 3$ (b) $5x^2 - 10x + 3y^2 - 12y = 4$
(c) $2x^2 - 4x + y^2 + 2y = 9$ (d) $4x^2 + 3x + 5y^2 + 8y = 3$
6. Draw the graphs of the following equations.
(a) $x^2 - 10x + 4y^2 - 8y = 7$ (b) $5x^2 - 20x + 3y^2 + 18y = 20$
(c) $2x^2 + 16x + y^2 + 4y = 5$ (d) $4x^2 - 3x + 5y^2 + 4y = 2$
*7. If $A > 0$ and $C > 0$ and we use the completing the square technique to rewrite the equation $Ax^2 + Bx + Cy^2 + Dy = E$ as in the examples, the left side of the resulting equation is the sum of two squares and therefore will be nonnegative for all possible values of x and y. How large must E be so that there are points satisfying the equation, that is, so that the right side is nonnegative?

*11. Parabolas

The graph of the equation $y = x^2$ is one of a family of curves called *parabolas*. Figure 6-58 provides the graph of this equation and three others. The tables of values for these equations illustrate how the Y-coordinates for the other three can be obtained from these for $y = x^2$ just by multiplying the Y-coordinate by the coefficient of x^2.

Figure 6–58

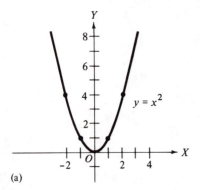

(a)

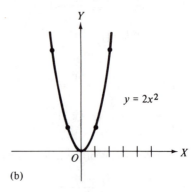

(b)

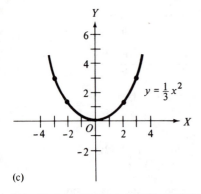

(c)

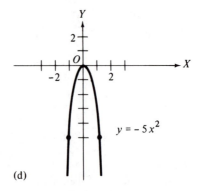

(d)

Figure 6–59

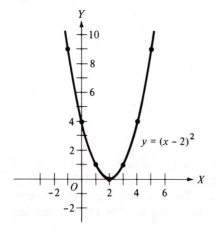

Example 1

x	-4	-3	-2	-1	0	1	2	3	4
y	16	9	4	1	0	1	4	9	16

$y = x^2$

x	-4	-3	-2	-1	0	1	2	3	4
y	32	18	8	2	0	2	8	18	32

$y = 2x^2$

x	-4	-3	-2	-1	0	1	2	3	4
y	$\frac{16}{3}$	3	$\frac{4}{3}$	$\frac{1}{3}$	0	$\frac{1}{3}$	$\frac{4}{3}$	3	$\frac{16}{3}$

$y = \frac{1}{3}x^2$

x	-4	-3	-2	-1	0	1	2	3	4
y	-80	-45	-20	-5	0	-5	-20	-45	-80

$y = -5x^2$ □

This multiplication can be done mentally once you become familiar with the Y-coordinates of $y = x^2$.

Example 2

Consider now the equation

$$y = (x - 2)^2$$

and the table of values below:

x	-2	-1	0	1	2	3	4	5	6
y	16	9	4	1	0	1	4	9	16

Plotting these and sketching the curve yields the graph given in Figure 6-59. Notice that this curve is shaped just like the graph of $y = x^2$; it is just shifted 2 units to the right. The "point" or *vertex* is at $(2, 0)$ instead of at $(0, 0)$ as with the graph of $y = x^2$ and the others in Figure 6-58. □

Example 3

Consider also the graph of $y = (x + 3)^2$ as given in Figure 6-60. This has the same shape as the graph of $y = x^2$ but is moved 3 units to the left. □

Figure 6-60

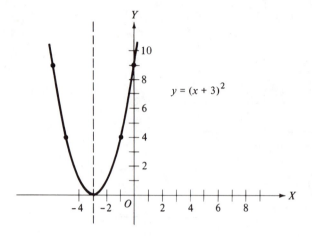

In general, the graph of

$$y = (x - h)^2$$

has the same shape as the graph of $y = x^2$. It is a parabola with its lowest point at $(h, 0)$; the graph is moved to the right if $h > 0$ and to the left if $h < 0$.

Example 4

The graph of $y = (x + \frac{7}{2})^2$ is given in Figure 6-61(a). □

Figure 6-61

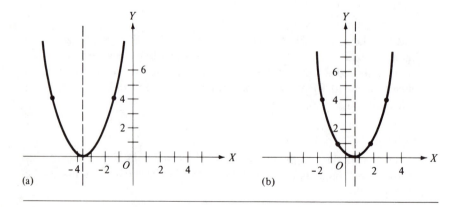

(a)   (b)

Example 5

The graph of $y = (x - \frac{3}{4})^2$ is given in Figure 6-61(b). □

Example 6

Just as the graph of $y = 3x^2$ can be drawn by multiplying each Y-coordinate of the graph of $y = x^2$ by 3, the graph of

$$y = 3(x - 2)^2$$

can be drawn by multiplying each Y-coordinate of the graph of $y = (x - 2)^2$ by 3. See Figure 6-62. □

Figure 6-62

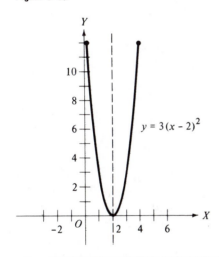

$y = 3(x - 2)^2$

Example 7

The graph of $y = x^2 + 2$ has the same shape as the graph of $y = x^2$ except that each Y-coordinate is moved up by 2. See Figure 6-63. □

Example 8

Consider now the equation

$$y = x^2 - 4x + 5$$

This does not fall into any of the patterns exemplified above but we will transform it into a form for which we can use previous knowledge in our graphing.

Figure 6–63

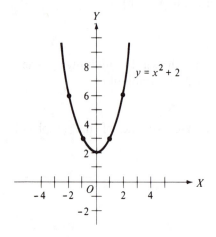

$y = x^2 - 4x + 5$
$y = x^2 - 4x + 4 + 1$
$y = (x - 2)^2 + 1$

or

$$y - 1 = (x - 2)^2$$

This equation has the same shape as the graph of $y = (x - 2)^2$ but the Y-coordinates have all been moved up by 1. The graph of $y = (x - 2)^2$ has the same shape as the graph of $y = x^2$ but is shifted 2 units to the right. Thus if we draw dotted horizontal and vertical lines through the point $(2, 1)$ and then draw the graph of $y = x^2$ on these dotted axes, we will have drawn the graph of $y = x^2 - 4x + 5$. Notice that the vertex is at $(2, 1)$. See Figure 6-64. $\square$

Example 9

To draw the graph of

$$y = 2x^2 + 12x + 7$$

we again employ the "completing the square" technique.

$$
\begin{aligned}
y - 7 &= 2x^2 + 12x \\
&= 2(x^2 + 6x) \\
y - 7 + 18 &= 2(x^2 + 6x + 9) \\
y + 11 &= 2(x + 3)^2
\end{aligned}
$$

Figure 6-64

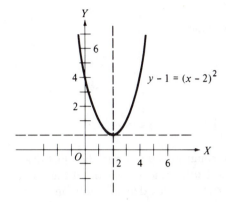

When $x = -3$, the right side is 0 and so y must be -11. Thus the vertex is at $(-3, -11)$. This equation has a graph with the same shape as that of $y = 2x^2$ but it is shifted 3 to the left and down 11; that is, if we draw our "new dotted axes" through $(-3, -11)$, we then draw the graph of $y = 2x^2 + 12x + 7$ by drawing the graph of $y = 2x^2$ on the "new axes." See Figure 6-65. $\square$

Figure 6-65

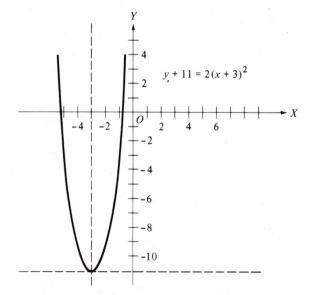

In general, the equation

$$y = Ax^2 + Bx + C \qquad \text{with } A \neq 0$$

can be transformed by completing the square into

$$y - \frac{4AC - B^2}{4A} = A\left(x + \frac{B}{2A}\right)^2$$

The vertex will be at the point

$$\left(\frac{-B}{2A}, \frac{4AC - B^2}{4A}\right) = \left(\frac{-B}{2A}, C - \frac{B^2}{4A}\right)$$

We can draw the "new dotted axes" through this point and then just draw the graph of $y = Ax^2$ on the new axes. Notice that the shape of the curve depends only on A, the X-coordinate of the vertex depends only on A and B and the Y-coordinate of the vertex depends on A, B, and C.

Example 10

The graph of $y = 2x^2 + 4x + 3$ has the vertex at $(-1, 1)$ and the same shape as $y = 2x^2$. See Figure 6-66. ☐

Example 11

The graph of $y = -3x^2 - 6x + 5$ has the vertex at $(-1, 8)$ and has the same shape as $y = -3x^2$. We draw it by drawing our new "dotted axes" and then drawing the graph of $y = -3x^2$ on the new axes. See Figure 6-67. ☐

Figure 6-66

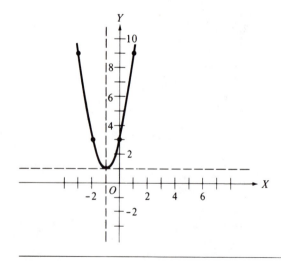

Figure 6-67

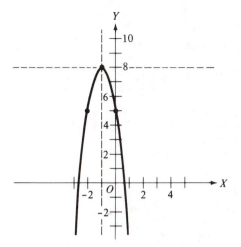

In general, you do not need to remember the entire formula for the coordinates of the vertex. If you know that the X-coordinate is $-B/2A$, then you can compute the Y-coordinate by substituting $-B/2A$ for x in the equation.

EXERCISES

1. Draw the graphs of the following equations.
   (a) $y = 4x^2$ (b) $y = -2x^2$
   (c) $y = \frac{1}{2}x^2$ (d) $y = \frac{2}{3}x^2$
   (e) $y = -\frac{4}{3}x^2$ (f) $y = 6x^2$

2. Draw the graphs of the following equations.
   (a) $y = 3x^2$ (b) $y = -4x^2$
   (c) $y = \frac{1}{3}x^2$ (d) $y = \frac{3}{4}x^2$
   (e) $y = -\frac{3}{2}x^2$ (f) $y = 5x^2$

3. Draw the graphs of the following equations.
   (a) $y = 4x^2 + 2$ (b) $y = -2x^2 - 3$
   (c) $y = \frac{1}{2}x^2 + \frac{2}{3}$ (d) $y = \frac{2}{3}x^2 + 4$
   (e) $y = -\frac{4}{3}x^2 - 5$ (f) $y = 6x^2 - 3$

4. Draw the graphs of the following equations.
   (a) $y = 3x^2 + 4$ (b) $y = -4x^2 - 2$
   (c) $y = \frac{1}{3}x^2 - \frac{1}{4}$ (d) $y = \frac{3}{4}x^2 - 5$
   (e) $y = -\frac{3}{2}x^2 + 2$ (f) $y = 5x^2 - 9$

5. Draw the graphs of the following equations.
   (a) $y = 4(x - 1)^2$ (b) $y = -2(x + 2)^2$
   (c) $y = \frac{1}{2}(x - 3)^2$ (d) $y = \frac{2}{3}(x - 2)^2$
   (e) $y = -\frac{4}{3}(x - 4)^2$ (f) $y = 6(x + \frac{1}{2})^2$

6. Draw the graphs of the following equations.
   (a) $y = 3(x + 1)^2$
   (b) $y = -4(x - 2)^2$
   (c) $y = \frac{1}{3}(x + 5)^2$
   (d) $y = \frac{3}{4}(x + 3)^2$
   (e) $y = -\frac{3}{2}(x - 5)^2$
   (f) $y = 5(x - \frac{1}{3})^2$

7. Draw the graphs of the following equations.
   (a) $y = 4(x - 3)^2 + 2$
   (b) $y = -2(x + 4)^2 - 3$
   (c) $y = \frac{1}{2}(x + 1)^2 + \frac{2}{3}$
   (d) $\frac{2}{3}(x - 1)^2 + \frac{2}{3} = y$
   (e) $y = -\frac{4}{3}(x + 3)^2 - 5$
   (f) $y = 6(x - \frac{1}{3})^2 + \frac{3}{4}$

8. Draw the graphs of the following equations.
   (a) $y = 3(x + 2)^2 + 1$
   (b) $y = -4(x - 3)^2 - 2$
   (c) $y = \frac{1}{3}(x - 6)^2 + 3$
   (d) $y = \frac{3}{4}(x - 4)^2 + 2$
   (e) $y = -\frac{3}{2}(x + 7)^2 - 5$
   (f) $y = 5(x + \frac{1}{2})^2 - \frac{4}{3}$

9. Draw the graphs of the following equations.
   (a) $y = 4x^2 + 8x - 5$
   (b) $y = -2x^2 + 16x - 2$
   (c) $y = \frac{1}{2}x^2 + 3x + 2$
   (d) $y = \frac{2}{3}x^2 + 2x + 1$
   (e) $y = -\frac{4}{3}x^2 + 8x - 5$
   (f) $y = 6x^2 + 2x - 1$

10. Draw the graphs of the following equations.
    (a) $y = 3x^2 + 6x + 2$
    (b) $y = -4x^2 + 24x - 7$
    (c) $y = \frac{1}{3}x^2 + 2x + 5$
    (d) $y = \frac{3}{4}x^2 + 3x + 2$
    (e) $y = -\frac{3}{2}x^2 - 6x - 5$
    (f) $y = 5x^2 + 3x - 4$

11. Draw the graphs of the following equations.
    (a) $y = 2x^2 + 8x - 3$
    (b) $y = x^2 + 6x - 7$
    (c) $y = 4x^2 - 12x + 3$
    (d) $y = -3x^2 + 4x - 2$
    (e) $y = -5x^2 + 30x - 4$
    (f) $y = \frac{1}{3}x^2 + \frac{1}{2}x - 2$

12. Draw the graphs of the following equations.
    (a) $y = x^2 + 8x - 7$
    (b) $y = 3x^2 + 4x + 5$
    (c) $y = -2x^2 + 5x + 3$
    (d) $y = 3x^2 + 8x - 9$
    (e) $y = 7x^2 + 11x - 2$
    (f) $y = \frac{3}{4}x^2 - \frac{1}{3}x - 4$

*13. (a) At what points does the graph of $y = Ax^2 + Bx + C$ cross the X-axis if, in fact, it does so?

(b) If the above graph does not cross the X-axis, what can be said about A, B, and C? Justify your answer.

(c) If the above graph does not cross the X-axis and $A > 0$, where is the graph located? If $A < 0$, where is it? Justify your answer.

*12. Hyperbolas

Graphs of equations of the form

$$\frac{x^2}{a^2} - \frac{y^2}{b^2} = 1$$

are among a family of curves called *hyperbolas*.

Example 1

The graph of

$$\frac{x^2}{9} - \frac{y^2}{4} = 1$$

is a hyperbola. We can draw this by making the table below and using the "plot and sketch" method.

x	± 3	± 5	± 6	± 7
y	0	$\pm\frac{8}{3}$	$\pm 2\sqrt{3}$	$\pm\frac{4}{3}\sqrt{10}$

This table actually provides us with 14 points to use as starting points. (See Figure 6-68.) ☐

There is a simpler method for drawing such graphs. Consider the equation above:

$$\frac{x^2}{9} - \frac{y^2}{4} = 1$$

and the equation

$$\frac{x^2}{9} - \frac{y^2}{4} = 0$$

Figure 6-68

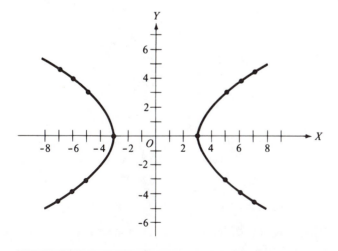

The first equation is equivalent to

$$\frac{y^2}{4} = \frac{x^2}{9} - 1$$

or

$$\frac{y^2}{4} = \frac{x^2 - 9}{9}$$

$$y^2 = \tfrac{4}{9}(x^2 - 9)$$

and, therefore,

$$y = \pm\tfrac{2}{3}\sqrt{x^2 - 9}$$

The second equation is equivalent to

$$\frac{y^2}{4} = \frac{x^2}{9}$$

or

$$y^2 = \tfrac{4}{9}x^2$$

and, therefore,

$$y = \pm\tfrac{2}{3}\sqrt{x^2}$$
$$y = \pm\tfrac{2}{3}|x|$$

that is,

$$y = \tfrac{2}{3}x \quad \text{or} \quad y = -\tfrac{2}{3}x$$

Thus the graph of

$$\frac{x^2}{9} - \frac{y^2}{4} = 1$$

is the graph of

$$y = \pm\tfrac{2}{3}\sqrt{x^2 - 9}$$

The graph of

$$\frac{x^2}{9} - \frac{y^2}{4} = 0$$

is the graph of

$$y = \tfrac{2}{3}x \quad \text{or} \quad y = -\tfrac{2}{3}x \quad \text{or just} \quad y = \pm\tfrac{2}{3}|x|$$

The graph of $y = \pm\tfrac{2}{3}|x|$ is a pair of lines meeting at the origin, one with slope $\tfrac{2}{3}$ and the other with slope $-\tfrac{2}{3}$.

For $|x| \geq 3$,

$$\sqrt{x^2 - 9} < \sqrt{x^2} = |x|$$

Thus

$$\tfrac{2}{3}\sqrt{x^2 - 9} < \tfrac{2}{3}|x|$$

and

$$-\tfrac{2}{3}\sqrt{x^2 - 9} > -\tfrac{2}{3}|x|$$

Hence the curve must lie below the graph of $y = \tfrac{2}{3}|x|$ as in Figure 6-69 and above the graph of $y = -\tfrac{2}{3}|x|$ as in Figure 6-70. Furthermore, as $|x|$ is chosen larger, $\sqrt{x^2 - 9}$ gets closer to $\sqrt{x^2} = |x|$ and the curve gets close to the line.

Notice that there are no points on the curve for which $|x| < 3$ since if $|x| < 3$, $x^2 < 9$ and $\sqrt{x^2 - 9}$ has no meaning.

Thus to draw the graph of

$$\frac{x^2}{9} - \frac{y^2}{4} = 1$$

(i) We draw the graph of $|y| = \tfrac{2}{3}|x|$ or just the lines $y = \tfrac{2}{3}x$ and $y = -\tfrac{2}{3}x$.

(ii) We plot the points $(-3, 0)$ and $(3, 0)$ on the X-axis.

(iii) We draw a smooth curve from $(3, 0)$ upward which approaches the line $y = \tfrac{2}{3}x$ as x grows larger and draw one from $(3, 0)$ downward which approaches the line $y = -\tfrac{2}{3}x$ as x becomes larger. These two curves have the same shape.

(iv) In a similar manner we draw smooth curves from $(-3, 0)$ approaching these same lines.

Figure 6-69

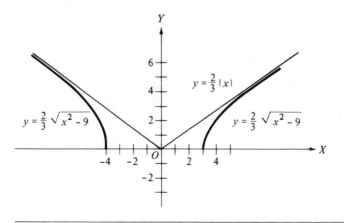

Figure 6–70

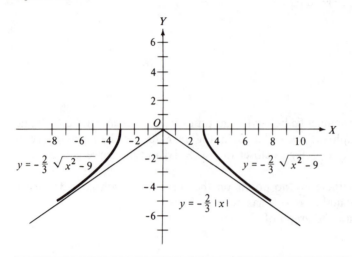

Of course, there is nothing special about the numbers 9 and 4 or 3 and 2. By generalizing the above argument we can easily show that an equation of the form

$$\frac{x^2}{a^2} - \frac{y^2}{b^2} = 1 \qquad \text{with } a > 0 \text{ and } b > 0$$

can be transformed into

$$y^2 = \frac{b^2}{a^2}(x^2 - a^2)$$

or

$$y = \pm \frac{b}{a}\sqrt{x^2 - a^2}$$

This curve will approach the lines

$$y = \frac{b}{a}x \quad \text{and} \quad y = \frac{-b}{a}x$$

as $|x|$ is chosen larger since

$$\left|\frac{b}{a}\sqrt{x^2 - a^2}\right| < \left|\frac{b}{a}x\right| \qquad \text{for } |x| > a$$

but $\sqrt{x^2 - a^2}$ gets close to $|x|$ as $|x|$ gets larger. These lines are called the *asymptotes* of the curve. Thus to draw the graph of an equation of the form

$$\frac{x^2}{a^2} - \frac{y^2}{b^2} = 1$$

(i) Draw the lines $y = (b/a)x$ and $y = (-b/a)x$. It is customary to draw these asymptotes as dotted lines.
(ii) Plot the points $(-a, 0)$ and $(a, 0)$.
(iii) Draw the smooth curves from these points approaching the asymptotes as $|x|$ gets larger. The curve will be below $y = (b/a)x$ and above $y = (-b/a)x$.

Such a hyperbola is said to have its *center* at $(0, 0)$ and the points $(-a, 0)$ and $(a, 0)$ are its *vertices*.

Example 2

The graph of $\dfrac{x^2}{16} - \dfrac{y^2}{9} = 1$ is given in Figure 6-71. □

Figure 6-71

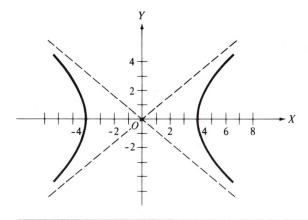

Example 3

The graph of $\dfrac{x^2}{1} - \dfrac{y^2}{\frac{1}{4}} = 1$ is given in Figure 6-72. □

Figure 6-72

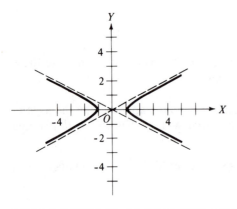

Example 4

The graph of $\dfrac{x^2}{4} - \dfrac{y^2}{16} = 1$ is given in Figure 6-73. $\square$

An equation of the form

$$\frac{y^2}{c^2} - \frac{x^2}{d^2} = 1$$

also has a graph which is a hyperbola. In this case the curve opens up and down instead of left and right. The vertices are on the Y-axis at $(0, -c)$

Figure 6-73

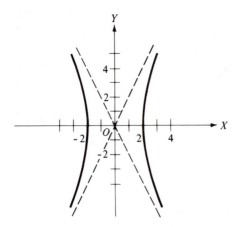

and $(0, c)$ and the asymptotes are the lines

$$y = \frac{c}{d}x \quad \text{and} \quad y = \frac{-c}{d}x$$

In this case the curves lie above the graph of $y = (c/d)|x|$ and below $y = (-c/d)|x|$. The justifications are just about the same as those given before and we omit them.

To draw the graph of

$$\frac{y^2}{c^2} - \frac{x^2}{d^2} = 1$$

(i) Draw the asymptotes $y = (c/d)x$ and $y = (-c/d)x$.
(ii) Plot $(0, -c)$ and $(0, c)$.
(iii) Draw smooth curves from these points approaching the asymptotes as x gets larger.

The points $(0, -c)$ and $(0, c)$ are called vertices and the center is at the origin.

Example 5

The graph of $\dfrac{y^2}{25} - \dfrac{x^2}{16} = 1$ is given in Figure 6-74. ☐

Figure 6-74

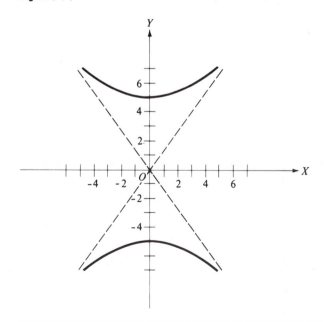

Figure 6-75

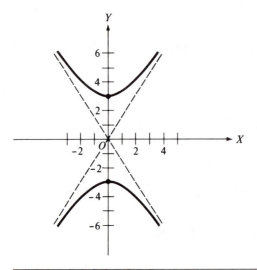

Example 6

The graph of $\dfrac{y^2}{9} - \dfrac{x^2}{4} = 1$ is given in Figure 6-75. ☐

Example 7

Consider now the equation

$$\frac{(x-5)^2}{9} - \frac{(y-7)^2}{4} = 1$$

When $y = 7$, $x - 5 = \pm 3$ and $x = 8$ or $x = 2$. Furthermore, $|x - 5| \geq 3$ or otherwise the left side is less than 1. If we compute a few points and sketch the curve we see that it has the same shape as

$$\frac{x^2}{9} - \frac{y^2}{4} = 1$$

It is just located in a different place in the plane.

If we draw dotted vertical and horizontal lines through $(5, 7)$ we see that the graph of

$$\frac{(x-5)^2}{9} - \frac{(7-7)^2}{4} = 1$$

is just the graph of

$$\frac{x^2}{9} - \frac{y^2}{4} = 1$$

drawn using these dotted lines as the axes. See Figure 6-76. □

Figure 6–76

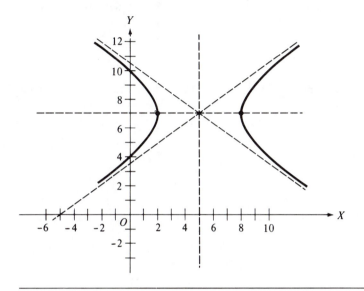

If we can draw graphs of equations of the form

$$\frac{x^2}{a^2} - \frac{y^2}{b^2} = 1$$

then we can draw graphs of equations of the form

$$\frac{(x - h)^2}{a^2} - \frac{(y - k)^2}{b^2} = 1$$

All we need do is draw a new dotted set of axes with the origin at (h, k) and draw the graph of

$$\frac{x^2}{a^2} - \frac{y^2}{b^2} = 1$$

on this "new" set of axes. The points $(h - a, k)$ and $(h + a, k)$ will be the vertices and (h, k) will be the center.

Similarly, the graph of an equation of the form

$$\frac{(y-k)^2}{c^2} - \frac{(x-h)^2}{d^2} = 1$$

will have the same shape as the graph of

$$\frac{y^2}{c^2} - \frac{x^2}{d^2} = 1$$

To draw this graph draw a new dotted set of axes with the origin at (h, k) and draw the graph of

$$\frac{y^2}{c^2} - \frac{x^2}{d^2} = 1$$

on this set of axes. The center will be at (h, k) and the vertices will be $(h, k - c)$ and $(h, k + c)$.

The completing the square technique can be used to transform an equation of the form

$$Ax^2 + Bx + Cy^2 + Dy = E \qquad \text{with } AC < 0$$

into one of the form

$$\frac{(x-h)^2}{a^2} - \frac{(y-k)^2}{b^2} = 1$$

$$\frac{(y-k)^2}{c^2} - \frac{(x-h)^2}{d^2} = 1$$

or

$$\frac{(x-h)^2}{a^2} - \frac{(y-k)^2}{b^2} = 0$$

In the last case, the graph is two lines meeting at (h, k).

Example 8

$$4x^2 + 16x - 25y^2 - 50y = 109$$
$$4(x^2 + 4x) - 25(y^2 + 2y) = 109$$
$$4(x^2 + 4x + 4) - 25(y^2 + 2y + 1) = 109 + 16 - 25$$
$$4(x + 2)^2 - 25(y + 1)^2 = 100$$
$$\frac{(x + 2)^2}{25} - \frac{(y + 1)^2}{4} = 1$$

See Figure 6-77. □

Figure 6-77

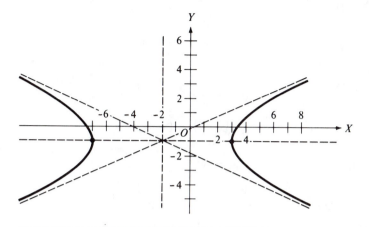

EXERCISES

1. Draw the graphs of the following equations.

(a) $\dfrac{x^2}{4} - \dfrac{y^2}{25} = 1$ (b) $\dfrac{x^2}{16} - \dfrac{y^2}{49} = 1$

(c) $\dfrac{x^2}{1} - \dfrac{y^2}{16} = 1$ (d) $\dfrac{x^2}{25} - \dfrac{y^2}{9} = 1$

(e) $\dfrac{y^2}{49} - \dfrac{x^2}{4} = 1$ (f) $\dfrac{y^2}{36} - \dfrac{x^2}{16} = 1$

(g) $\dfrac{y^2}{9} - \dfrac{x^2}{5} = 1$ (h) $\dfrac{y^2}{64} - \dfrac{x^2}{36} = 1$

2. Draw the graphs of the following equations.

(a) $\dfrac{x^2}{9} - \dfrac{y^2}{16} = 1$ (b) $\dfrac{x^2}{1} - \dfrac{y^2}{25} = 1$

(c) $\dfrac{x^2}{49} - \dfrac{y^2}{25} = 1$ (d) $\dfrac{x^2}{16} - \dfrac{y^2}{25} = 1$

(e) $\dfrac{y^2}{25} - \dfrac{x^2}{9} = 1$ (f) $\dfrac{y^2}{36} - \dfrac{x^2}{25} = 1$

(g) $\dfrac{y^2}{9} - \dfrac{x^2}{49} = 1$ (h) $\dfrac{y^2}{25} - \dfrac{x^2}{64} = 1$

3. Draw the graphs of the following equations.

(a) $\dfrac{(x-1)^2}{4} - \dfrac{(y-3)^2}{25} = 1$

(b) $\dfrac{(x+2)^2}{16} - \dfrac{(y-4)^2}{49} = 1$

(c) $\dfrac{(x+5)^2}{1} - \dfrac{(y-2)^2}{16} = 1$

(d) $\dfrac{(x-4)^2}{25} - \dfrac{(y+3)^2}{9} = 1$

(e) $\dfrac{(y-2)^2}{49} - \dfrac{(x-3)^2}{4} = 1$

(f) $\dfrac{(y+3)^2}{36} - \dfrac{(x-5)^2}{16} = 1$

(g) $\dfrac{(y-\frac{1}{2})^2}{9} - \dfrac{(x+2)^2}{5} = 1$

(h) $\dfrac{(y+5)^2}{64} - \dfrac{(x-2)^2}{36} = 1$

4. Draw the graphs of the following equations.

(a) $\dfrac{(x-2)^2}{9} - \dfrac{(y-5)^2}{16} = 1$

(b) $\dfrac{(x+3)^2}{1} - \dfrac{(y-2)^2}{25} = 1$

(c) $\dfrac{(x-5)^2}{49} - \dfrac{(y+4)^2}{25} = 1$

(d) $\dfrac{(x+7)^2}{16} - \dfrac{(y+3)^2}{25} = 1$

(e) $\dfrac{(y-4)^2}{25} - \dfrac{(x-2)^2}{9} = 1$

(f) $\dfrac{(y+5)^2}{36} - \dfrac{(x-3)^2}{25} = 1$

(g) $\dfrac{(y-\frac{1}{3})^2}{9} - \dfrac{(x+2)^2}{49} = 1$

(h) $\dfrac{(y+11)^2}{25} - \dfrac{(x-10)^2}{64} = 1$

5. Draw the graphs of the following equations.
  (a) $x^2 - 6x - 4y^2 + 8y = 11$
  (b) $2x^2 - 4x - y^2 + 2y = 3$
  (c) $5x^2 - 10x - 3y^2 - 12y = 22$
  (d) $4x^2 + 3x - 5y^2 + 10y = 3$
6. Draw the graphs of the following equations.
  (a) $x^2 - 8x - 4y^2 - 8y = 4$
  (b) $2x^2 - 8x - y^2 + 6y = 5$
  (c) $5x^2 + 20x - 3y^2 - 48y = 8$
  (d) $6x^2 + 8x - 5y^2 + 10y = 2$

Appendix

A Summary of the Arithmetic of Signed Numbers

Students are usually introduced to the arithmetic of signed numbers in junior high school mathematics. This appendix is included if you wish to refresh your memory of the techniques. The arithmetic of signed numbers is governed by the principles given in the axioms along with Lemmas 1 and 2 and Theorems 3 and 4 of Chapter 1. The basic principles are given in the axioms while the lemmas and theorems govern the relationships between products and sums involving negative numbers.

From Theorem 3 it follows that:

(i) The product of two positive numbers is the product of their absolute values and, therefore, is positive.

(ii) The product of two negative numbers is the product of their absolute values and, therefore, is positive.

(iii) The product of a positive number and a negative number is a negative number and, in fact, it is the negative of the product of their absolute values.

Examples

$$(7)(11) = 77$$
$$(-7)(-11) = 77$$
$$(7)(-11) = -77$$
$$(-7)(11) = -77 \quad \square$$

Thus, to multiply two signed numbers:

(a) Multiply their absolute values; this is the absolute value of the product.
(b) If both are positive or both are negative, the product is positive.
(c) If one is positive and the other negative, the product is negative.

To put it briefly:

(a) Multiply their absolute values.
(b) Affix a "+" sign if the numbers agree in sign and a "−" sign if they disagree in sign.

Dividing signed numbers follows much the same rules as multiplying them, because, after all, division is defined in terms of multiplication.
To divide one signed number by another:

(i) Divide the corresponding absolute values.
(ii) Affix a "+" sign if the original numbers agree in sign and a "−" sign if they disagree in sign.

Examples

$$\frac{-8}{-4} = +\left(\frac{8}{4}\right) = +2 = 2$$

$$\frac{-36}{9} = -\left(\frac{36}{9}\right) = -4$$

$$\frac{98}{-7} = -\left(\frac{98}{7}\right) = -14 \;\square$$

To add a positive number and a negative number:

(i) If they have the same absolute value, the sum is 0.
(ii) Subtract the smaller absolute value from the larger; this is the absolute value of the sum.
(iii) To this number affix the sign of the number with the larger absolute value.

From the discussion of subtraction we know that $a + (-b) = a - b$. Thus, if $a > b$, the difference we compute is just $a - b$, which is $a + (-b)$. But if $a < b$, the difference we compute is $b - a$ and $-(b - a) = a - b$ because

$$-(b - a) = -(b + (-a))$$
$$= (-b) + (-(-a))$$
$$= (-b) + a$$
$$= a + (-b)$$
$$= a - b$$

Examples

$$7 + (-5) = 7 - 5 = 2$$
$$8 + (-14) = -(14 - 8) = -6$$
$$(\tfrac{2}{3}) + (-\tfrac{5}{6}) = -(\tfrac{5}{6} - \tfrac{2}{3}) = -\tfrac{1}{6} \ \square$$

Subtracting signed numbers can frequently be simplified by using addition since

$$x - y = x + (-y) \qquad \text{for any numbers } x \text{ and } y$$

Examples

$$\begin{aligned}
18 - (-4) &= 18 + (-(-4)) \\
&= 18 + 4 \\
&= 22
\end{aligned}$$

$$\begin{aligned}
(-47) - 5 &= (-47) + (-5) \\
&= -(47 + 5) \\
&= -52
\end{aligned}$$

$$\begin{aligned}
(-38) - (-7) &= (-38) + (-(-7)) \\
&= (-38) + 7 \\
&= -31
\end{aligned}$$

$$\begin{aligned}
(-\tfrac{4}{5}) - (\tfrac{2}{3}) &= (-\tfrac{4}{5}) + (-\tfrac{2}{3}) \\
&= -(\tfrac{4}{5} + \tfrac{2}{3}) \\
&= -\tfrac{22}{15} \ \square
\end{aligned}$$

EXERCISES

1. Compute.
   (a) $(-59) + (-92)$
   (b) $38 + (-56)$
   (c) $(-131) + (43)$
   (d) $7.7 + (-3.2)$
   (e) $(-\tfrac{5}{6}) + (-\tfrac{1}{4})$
   (f) $\tfrac{2}{7} - (-98)$
   (g) $(-42) - (76)$
   (h) $(-19.2) - (-3.4)$
   (i) $(-47.9) - (32.3)$
   (j) $(\tfrac{7}{12}) - (-\tfrac{9}{20})$

2. Compute.
   (a) $(-95) + (-32)$
   (b) $47 + (-84)$
   (c) $(-242) + (57)$
   (d) $12.5 + (-6.2)$
   (e) $(-\tfrac{3}{4}) + (-\tfrac{2}{3})$
   (f) $541 - (-312)$
   (g) $(-93) - (41)$
   (h) $(-31.6) - (-23.7)$
   (i) $(-6.45) - (11.5)$
   (j) $(\tfrac{13}{15}) - (-\tfrac{3}{20})$

3. Compute.
   (a) $(43.)(-6.4)$
   (b) $(-57.2)(-4.4)$
   (c) $(84)(-65)$
   (d) $(-113)(.73)$
   (e) $(-.042)(-6.64)$
   (f) $(-4.7)(\tfrac{21}{32})$
   (g) $(-276)/14$
   (h) $1036/(-28)$
   (i) $(-125)/(-16)$
   (j) $(-525)/15$
   (k) $(-396)/(-32)$
   (l) $(-\tfrac{15}{28})/(-\tfrac{10}{7})$

4. Compute.
   (a) $(5.6)(-8.4)$
   (b) $(-61.7)(-6.4)$
   (c) $(97)(-76)$
   (d) $(-135)(.92)$
   (e) $(-.062)(-8.14)$
   (f) $(\frac{5}{18})(-\frac{12}{35})$
   (g) $(-703)/19$
   (h) $1032/(-43)$
   (i) $(-17.2)/(-32)$
   (j) $(-21)/1.5$
   (k) $(-4.26)/(-64)$
   (l) $(-\frac{7}{12})/(-\frac{35}{78})$

Answers to Odd-Numbered Exercises

CHAPTER 1

Section 1

1. $2W + 2L = P$
   where W is the width
   L is the length
   P is the perimeter
3. $A = \frac{1}{2}(a + b)h$
   A is the area
   a is the top
   b is the base
   h is the height
5. $A = \pi r^2$
   A is the area
   r is the radius
7. $v = \pi r^2 h$
   v is volume
   r is radius of base
   h is the height
9. $v = \frac{1}{3}\pi r^2 h$
   v is volume
   r is the radius of base
   h is the height

Section 2

1. (a) $\{a, b, c, d, e, f, g, h, k\}$ (b) $\{a, c, e\}$ (c) $\{b, d, f\}$ (d) $\{g, h, k\}$
   (e) $\{b, d, f\}$ (f) $\{b, d, f, g, h, k\}$
5. (a) T   (b) T   (c) T   (d) T   (e) T   (f) T
   (g) F   (h) T   (i) T   (j) F   (k) T   (l) T

Section 3

1. (a) T   (b) T   (c) F   (d) F   (e) T   (f) F

Section 4

1. (a) F   (b) F   (c) F   (d) T   (e) F   (f) T   (g) T
   (h) F   (i) T   (j) T   (k) T   (l) T   (m) T   (n) T

Section 5

1. yes
3. 1021 is prime
5. (a) $\frac{3}{7}$ (b) $\frac{4}{15}$ (c) $\frac{11}{7}$ (d) $\frac{4}{25}$
   (e) $\frac{49}{27}$ (f) $\frac{4}{3}$ (g) $\frac{5}{7}$ (h) $\frac{10}{13}$
7. In general if p has no divisor d other than 1, such that $d \cdot d < p$, p is prime.

CHAPTER 2

Section 1

1. (a) $x = 8.6$ (b) $y = 103$ (c) $z = 1.46$ (d) $w = -\frac{1}{12}$
   (e) $x = \frac{91}{45}$ (f) $y = 19$ (g) $z = 475$ (h) $x = 24$
   (i) $y = -\frac{9}{2}$ (j) $z = -23$
3. (a) $x = 24.125$ (b) $y = \frac{15}{28}$ (c) $x = \frac{8}{5}$ (d) $y = -\frac{71}{154}$
   (e) $x = 323\frac{1}{3}$ (f) $y = \frac{3}{94000}$ (g) $y = \frac{1325}{221}$ (h) $x = \frac{1598}{273}$
   (i) $x = \frac{98}{1581}$ (j) $y = -\frac{1483}{357}$

Section 2

1. (a) 2   (b) 6   (c) 1/2   (d) 8   (e) 8
   (f) -14 (g) $-\frac{5}{2}$ (h) $15\frac{1}{4}$ (i) $\frac{1}{4}$ (j) $-\frac{1}{2}$
3. (a) $-\frac{1713}{118}$ (b) 1   (c) $\frac{21}{22}$ (d) $-\frac{385}{153}$ (e) $\frac{58}{13}$ (f) $\frac{805}{54}$
5. no solution, since if so, $0 = 2$, which is a contradiction

Section 3

1. (a) 236.5 miles   (b) $r = \dfrac{d}{t}$ (c) $\frac{55}{7}$ yd/sec   (d) 450 mph

(e) $t = \dfrac{d}{r}$ (f) approximately 6 sec   (g) 36 sec
(h) approximately 1 sec

3. (a) 14.6 in.   (b) 6.5 cm   (c) 2.5 in.

5. (a) 54 cm^2 (b) $h = \dfrac{2A}{b_1 + b_2}$ (c) 22 in.   (d) 13 cm

7. $x = \dfrac{4 - 6y}{3} = \dfrac{4}{3} - 2y$

   if $y = 3$, $x = -\frac{14}{3}$

9. $x = \dfrac{5 - 13y}{11}$, $y = \dfrac{5 - 11x}{13}$

   if $x = 5$, $y = -\frac{50}{13}$
   if $y = 11$, $x = -\frac{138}{11}$

11. $c = \dfrac{2A}{r}$, $r = \dfrac{2A}{c}$

13. $h = \dfrac{A}{B}$

15. $y = \dfrac{c - ax}{b}$, $x = \dfrac{c - by}{a}$

17. $x = \dfrac{11 + 6y - 10z}{2}$, $y = \dfrac{2x + 10z - 11}{6}$, $z = \dfrac{11 - 2x + 6y}{10}$

Section 4

1. (a) $x = 4$, $y = 2$ (b) $x = \frac{112}{29}$, $y = -\frac{64}{29}$
   (c) $x = \frac{442}{39}$, $y = \frac{208}{39}$ (d) $x = \frac{162}{55}$, $y = -\frac{9}{110}$
   (e) $x = -6$, $y = 12$ (f) $x = 5$, $y = 7$
   (g) $x = \frac{17}{3}$, $y = -\frac{23}{3}$ (h) $x = \frac{398}{197}$, $y = -\frac{1949}{1970}$
   (i) $x = \frac{4942}{6039}$, $y = -\frac{56}{2013}$ (j) $x = \frac{469}{82}$, $y = -\frac{157}{82}$

3. (a) $x = 4$ (b) $x = -\frac{794}{347}$ (c) $x = \frac{26}{7}$ (d) $x = \frac{770}{69}$
   $y = 3$ $y = \frac{1635}{347}$ $y = \frac{5}{14}$ $y = -\frac{3560}{69}$
   (e) $x = 1$ (f) $x = -\frac{8}{17}$ (g) $x = -\frac{46410}{1969}$ (h) $x = \frac{23}{8}$
   $y = -1$ $y = -\frac{14}{17}$ $y = -\frac{85183}{1969}$ $y = -\frac{49}{8}$
   (i) $x = \frac{145}{44}$ (j) $x = 40$ (k) $x = \frac{143}{355}$ (l) $x = \frac{305}{127}$
   $y = \frac{37}{44}$ $y = 17$ $y = \frac{79}{355}$ $y = -\frac{169}{127}$
   (m) $x = \frac{335}{56}$ (n) $x = \frac{125}{96}$
   $y = -\frac{195}{28}$ $y = -\frac{149}{32}$

5. (a) $x = -52$ (b) $x = \frac{53}{31}$ (c) $x = \frac{87}{47}$ (d) $x = \frac{14}{13}$
   $y = 16$ $y = \frac{60}{31}$ $y = -\frac{1}{47}$ $y = \frac{4}{13}$

Section 5

1. $x = -4$ 3. no solution   5. $x = -\frac{1424}{265}$ 7. $x = 2$
   $y = 7$ $y = \frac{308}{53}$ $y = -2$
   $z = 0$ $z = \frac{262}{53}$ $z = 5$

9. $x = \frac{17259}{6496}$ 　　11. $x = 4$

$y = \frac{222}{203}$ 　　　　$y = -3$

$z = -\frac{205}{1624}$ 　　　$z = 2$

*13. $x = \dfrac{10p + 9q - 6r}{47}$

$y = \dfrac{9p - 6q + 4r}{47}$

$z = \dfrac{9r - 15p + 10q}{94}$

15. $x = 6$ 　　　17. $x = \frac{118}{13}$ 　　　21. $x = \frac{103}{33}$

$y = -2$ 　　　　$y = -\frac{96}{13}$ 　　　$y = -\frac{27}{11}$

$z = 1$ 　　　　　$z = \frac{59}{13}$ 　　　　$z = \frac{90}{11}$

　　　　　　　　　　　　　　　　　　$w = -\frac{14}{33}$

Section 6

1. (a) $x > -\frac{3}{2}$ 　　(b) $x < 4$ 　　(c) $x < -4$ 　　(d) $x \le 9$
   (e) $x \ge -\frac{11}{8}$ 　(f) $x > -\frac{7}{2}$ 　(g) $x \ge -\frac{5}{2}$ 　(h) $x > -\frac{235}{98}$
   (i) $y \le 2$ 　　　(j) $x < 26$ 　　(k) $x \le \frac{34}{47}$ 　(l) $x < \frac{59}{60}$
3. (a) $-\frac{3}{2} < x < 4$ 　　　(b) $-\frac{11}{8} \le x \le 9$ 　　(c) $\frac{1}{3} < x < 2$
   (d) $x < \frac{5}{3}$ 　　　　　(e) $x > \frac{1}{4}$ or $x < 1$ 　(f) $\frac{1}{2} < x < \frac{4}{3}$
   (g) $x \ge \frac{24}{13}$ or $x \le \frac{14}{13}$ 　(h) $x \ge \frac{25}{46}$ or $x \le \frac{20}{3}$ 　(i) $x > 5$
   (j) $x > 3$

5. (a) $-\frac{7}{3} < x < 3$ 　　(b) $-\frac{5}{4} \le x \le -\frac{1}{x}$ 　　(c) $-\frac{2}{7} < x < \frac{8}{7}$

   (d) $x < -\frac{5}{2}$ or $x > \frac{7}{2}$ 　(e) $-4 \le x \le \frac{16}{3}$ 　　(f) $-\frac{29}{15} \le x \le \frac{11}{5}$
   (g) $x \le \frac{3}{8}$ or $x \ge \frac{11}{8}$ 　(h) $x < -\frac{134}{91}$ or $x > -\frac{4}{7}$ 　(i) $-\frac{3}{28} < x < \frac{45}{28}$
   (j) $x \le -\frac{25}{24}$ or $x \ge -\frac{5}{8}$

Section 7

1. 9.5 and 4.8 　　3. $m = -2, b = 5$ 　　5. 2.55 　　7. $\frac{6}{19}$ gal
9. approximately 2.68 lb of peanuts and 2.44 lb of cashews, almonds, and pecans
11. 2 gallons 　　13. 66 hits (at least) 　　15. 7 people
17. \$55.84/mo 　　19. 56.36 sec (approx.) 　　21. about 2 hr 10 min
23. $7\frac{2}{3}' \times 38\frac{1}{3}'$
27. your speed $= \frac{500}{91}$ mph
    water speed $= \frac{150}{91}$ mph
29. (a) approx. \$3.75 　　(b) approx. 1 hr 34 min 　　(c) approx. \$2.41/hr
    approx. 45¢/gal
31. 2.4 gal
33. $K = 13000$
    $C = 680$

CHAPTER 3

Section 1

1. (a) 32   (b) $\frac{8}{27}$ (c) $\frac{1}{16}$ (d) 441   (e) $\frac{49}{9}$
  (f) 0.00001   (g) 0.000001   (h) 9216

3. (a) $2x^6 - 19x^5 + 48x^4 - 101x^3 + 138x^2 - 188x + 77$
  (b) $12x^6 + 7x^5 - 48x^4 + 138x^3 - 142x^2 + 61x - 18$
  (c) $28x^8 + 63x^7 - 45x^6 - 20x^5 + 45x^4 - 63x^3 - 71x^2 + 27x - 18$
  (d) $2x^3y^5 + 6x^5y^4 - 32x^4y^4 - 5x^3y^5 - 21x^5y^3 + 41x^4y^3 + 43x^3y^4 - 96x^3y^3$
      $- 45x^2y^4 - 21x^4y^2 + 45x^2y^3 + 63x^3y^2$
  (e) $2x^4y^5z^4 + 6x^3y^6z^4 + 3x^4y^4z^4 + 9x^3y^5z^4 + 5x^3y^5z^3 + 15x^2y^6z^3 + 10x^4y^3z^3$
      $+ 15x^4y^2z^3 + 25x^3y^3z^2$
  (f) $6x^7 - 41x^5 + 106x^4 - 137x^3 + 119x^2 - 100x + 42$
  (g) $20x^8 - 59x^7 + 95x^6 - 46x^5 - 60x^4 + 43x^3 + 113x^2 - 236x + 85$
  (h) $\frac{1}{3}x^5 - \frac{11}{36}x^4 + \frac{11}{90}x^3 - \frac{9}{40}x^2 + \frac{1}{300} + \frac{1}{20}$
  (i) $2.52x^7 - 22.26x^6 + 7.696x^5 + 7.964x^4 + 48.632x^3 + 18.552x^2$
      $- 38.88x - 25.56$
  (j) $828x^5 - 157x^4 - 3469x^3 - 1447x^2 + 2746x + 1748$

5. (a) $x^3 + 7x^2 + 7x + 17$ remainder 32   (b) $x^2 - 8x + 11$ remainder 6
  (c) $x^3 - x^2 - 3x + 6$ remainder $3x + 10$
  (d) $3x^5 + 3x^4 + \frac{11}{2}x^3 + \frac{13}{4}x^2 + \frac{25}{8}x + \frac{97}{16}$ remainder $\frac{693}{16}x + \frac{207}{16}$
  (e) $3x^3 - 4x^2 - 5x - 14$ remainder $-x + 23$
  (f) $.6x^2 + 4.18x - 1.335$ remainder $4.9252x + 1.0935$
  (g) $4x^2 - 3x + 6$ (h) $5x^2 - 20x + 75$ remainder $-404x + 462$
  (i) $x^3 + \frac{1}{2}x^2 + \frac{13}{4}x - \frac{31}{8}$ remainder $\frac{189}{8}x - \frac{133}{8}$
  (j) $x^2 + x + 1$ (k) $x^4 - x^3 + x^2 - x + 1$
  (l) $x^3 + 2x^2 + 4x + 8$ (m) $x^6 + x^5 + x^4 + x^3 + x^2 + x + 1$
  (n) $x^2 + xy + y^2$ (o) $x^3 + x^2y + xy^2 + y^3$
  (p) $a^4 + a^3b + a^2b^2 + ab^3 + b^4$ (q) $a^5 + a^4b + a^3b^2 + a^2b^3 + ab^4 + b^5$

Section 2

1. $10x^2 + 17x + 3$ 3. $21a^2 + 13ab + 2b^2$
5. $24x^2 - 26xy - 15y^2$ 7. $63x^2 - 18xz$
9. $10x^2 + 12x + 3$ 11. $2y^2 + 23yz + 30z^2$
13. $12x^2 + 19xy + 5y^2$ 15. $36a^2 - 33ab - 119b^2$
17. $15x^2 - 16x + 4$ 19. $6x^2 + 59xy + 143y^2$
21. $x^2 - 1$ 23. $9x^2 - y^2$
25. $4x^2 - 9y^2$ 27. $x^2 - 4$
29. $36x^2 - 49$ 31. $x^2 + 4xy + 4y^2$
33. $a^2 + 4ab + 4b^2$ 35. $49a^2 + 28ab + 4b^2$
37. $100x^2 + 60xy + 9y^2$ 39. $81x^2 + 36x + 4$

Section 3

1. (a) $4x^2y(y^2 - 2x)$ (b) $3x^2y^2(6xy^2 - 3y^3 + 10x^2)$
  (c) $5x^2y^2z^2(4yz^2 - 7xz + 9x^2y^2)$ (d) $49x^3(x^2 + 2 - 7x^4)$

3. (a) $(2a + 1)(3a - 4)$ (b) $(2y - 3)(3y + 2)$
 (c) $(5x + 6y)(x - y)$ (d) $(x + 4y)(x - 2y)$
 (e) $(5x + 3y)(2x - 3y)$ (f) $4x^3(x^2 - 2x - 1)$
 (g) $2xy(2x + 5y)(3x - 7y)$ (h) $3x^2y^2(x - y)(x + y)$
 (i) $(x - 1)(x - 27)$ (j) $(3x + 2)(5x + 4)$

5. (a) $3x(x - 2x + 35)$ (b) $(4x + 5y)(3x - y)$
 (c) $(7a - b)(5a + b)$ (d) $2(6x + 5y)(x + 11y)$
 (e) $3a^2(3a - 2)(5a + 2)$ (f) $(2x - 11)(3x + 13)$
 (g) $(10x + 3)(8x + 7)$ (h) $3(3x - 1)(9x - 2)$
 (i) $(10x + 7)(10x + 3)$ (j) $(7x + 3)(9x - 11)$

7. (a) $(x + 2y)(x^2 - 2xy + 4y^2)$ (b) $(a - 5b)(a^2 + 5ab + 25b^2)$
 (c) $(2y - 3)(4y^2 + 6y + 9)$ (d) $(y^2 + 4)(y + 2)(y - 2)$
 (e) $(3x + 1)(9x^2 - 3x + 1)$ (f) $4(x + 5)(x^2 + 5x + 25)$
 (g) $(y - 3)(y^4 + 3y^3 + 9y^2 + 27y + 81)$ (h) $2(a - 3b^2)(a^2 + 3ab^2 + 9b^4)$
 (i) $(2x + 7y)(4x^2 - 14xy + 49y^2)$ (j) $(10x + 1)(100x^2 - 10x + 1)$

Section 4

1. (a) $\frac{2}{3}$ (b) $\frac{5}{14}$ (c) $\frac{3}{5}$ (d) $\frac{1}{4}$ (e) $\frac{3}{5}$

3. (a) $\dfrac{a + b}{a - b}$ (b) $\dfrac{x - 1}{x + 2}$ (c) $\dfrac{x + 3}{x + 2}$

 (d) $\dfrac{x + 3}{x + 2}$ (e) $\dfrac{x + 3y}{x + y}$ (f) $\dfrac{x + y}{x^2 + xy + y^2}$

 (g) $\dfrac{x - 5y}{x - 3y}$ (h) $\dfrac{(x - 2y)(x + 2y)}{5(x^2 - 4xy - 4y^2)}$ (i) $\dfrac{(x^2 + 1)(x + 1)}{(x^2 + x - 1)}$

 (j) $\dfrac{x^4 + x^2 + 1}{x^2 + 1}$ (k) $\dfrac{2x - 5}{x - 4}$ (l) $\dfrac{2x + 7}{x(2x + 2)}$

Section 5

1. (a) $\dfrac{1}{3x}$ (b) $(x + 2)^2 = x^2 + 4x + 4$

 (c) $\dfrac{x + 3y}{3x + y}$ (d) $\frac{1}{3}$

 (e) $\dfrac{x - 1}{x + 1}$ (f) $\dfrac{2}{x(x + 1)} = \dfrac{2}{x^2 + x}$

 (g) $\dfrac{2(x + 1)}{x - 1} = \dfrac{2x + 2}{x - 1}$ (h) $(x - 2y)^2 = x^2 - 4xy + 4y^2$

 (i) $\dfrac{2x - 2}{3x - 15}$ (j) $\dfrac{5(x^2 + x + 1)}{3(x + 1)} = \dfrac{5x^2 + 5x + 5}{3x + 3}$

 (k) $\dfrac{a^3 - 10a^2b + 31ab^2 - 30b^2}{a^5 - 6a^4b - 200ab^2 - 16a^2b^2 - 200b^3}$

(l) $\dfrac{x(x+1)}{x^2+x+1} = \dfrac{x^2+x}{x^2+x+1}$

(m) $\dfrac{4x^5+12x^4+9x^3+4x^2+12x+9}{x^5+11x^4-12x^3+27x^2+297x-324}$

(n) $\dfrac{(x-1)(x+2)}{(x+1)^2} = \dfrac{x^2+x-2}{x^2+2x+1}$

3. (a) $\dfrac{x+3}{x(x-7)}$

(b) $(x-2)(x-3) = x^2-5x+6$

(c) $\dfrac{a^2+ab-6b^2}{a^2-ab-6b^2}$

(d) $\dfrac{x^3+x^2-x-1}{x^3+2x^2-4x-8}$

(e) $\dfrac{x^3-2x^2-40x-64}{x^3+4x^2-29x+24}$

(f) $\dfrac{x^4-29x^2+100}{x^2-4x+5}$

Section 6

1. (a) $\frac{3}{2}$ (b) $\frac{22}{21}$ (c) $\frac{11}{24}$ (d) $\frac{43}{80}$ (e) $-\frac{29}{21}$ (f) $\frac{23}{27}$
   (g) $\frac{125}{216}$ (h) $\frac{1887}{3290}$

3. (a) $\dfrac{2x^2}{x^2-1}$

(b) $\dfrac{1}{x+1}$

(c) $\dfrac{10x^2+19xy-y^2}{6x^2-xy-y^2}$

(d) $\dfrac{x^4+x^3-9x^2+43x-6}{x^4+3x^3+7x+81}$

(e) $\dfrac{3x+1}{2x}$

(f) $\dfrac{x^3-x^2-2x^2y+4y^2x-3xy-2y^2-8y^3}{x^2-xy-2y^2}$

(g) $\dfrac{5x^2-17x}{x^3-5x^2-x+5}$

(h) $\dfrac{5x^2+2x+2}{x^4+x^3-x-1}$

(i) $\dfrac{2x^3+8x^2-2x+15}{x^3+2x^2-9x-18}$

(j) $\dfrac{2x^3+3x^2y-23xy^2+15y^3}{x^3-4xy^2+5x^2y-20y^3}$

(k) $\dfrac{a^4+3a^3b+3a^2b^2-2b^3-a^3-4a^2b-4ab^2}{a^4-ab^3+a^3b-b^4}$

(l) $\dfrac{9a^2b+14ab^2-20b}{a^3+8a^2b+19ab^2+12b^3}$

5. (a) $\dfrac{x^2-1}{x^2+1}$ (b) $\dfrac{y}{x-y}$ (c) $\dfrac{x^2+2x-8}{x}$ (d) $\dfrac{a^2+2ab+b^2}{a^2+ab+b^2}$

Section 7

1. 2, 1, -68, 917
3. (a) -35, -11, -26, 10, 64, 40, 469, -35, 3514, 1610   (b) no
5. (a) -2, 38, -4, 28, 78, 6, 328, 8   (b) no
7. $(x+3)(x^2+3x+12)$ 9. $(x-2)(x-3)(x+3)(x^2+2x+4)$

CHAPTER 4

Section 1

1. (a) $\frac{1}{25}$ (b) 1   (c) 100   (d) .001   (e) $\frac{1}{16}$ (f) $\frac{1}{9}$
   (g) $\frac{25}{16}$ (h) $\frac{27}{8}$

3. (a) 547,000   (b) 0.00694   (c) 87,600,000
   (d) 0.0000000992   (e) 81,320,000,000,000   (f) 0.00006429
   (g) 0.0003126   (h) 0.07519

Section 2

1. (a) $\sqrt{35}$ (b) $\sqrt{152}$ (c) $\sqrt{39.56}$ (d) $\sqrt{55x^4 y^6}$

   (e) $\sqrt{3/7}$ (f) $\sqrt{5}$ (g) $\sqrt{\dfrac{162}{21}}$ (h) $\sqrt{\dfrac{5x^3 y^8}{21z^8}}$

3. (a) $4\sqrt{3}$ (b) $5\sqrt{2}$ (c) $3\sqrt{3}$
   (d) $3\sqrt{7}$ (e) $2x^2 y^2 \sqrt{2x}$ (f) $4a^4 b^3 c^4 \sqrt{2a}$
   (g) $11(x+2y)^3 \sqrt{2(x+2y)}$ (h) $2x^3 y^4 z^4 \sqrt{17xz}$ (i) $3xy^3$
   (j) $25xy^2$

5. (a) $\dfrac{\sqrt{7}}{3}$ (b) $\dfrac{\sqrt{11}}{7}$ (c) $\dfrac{\sqrt{15}}{5}$ (d) $\dfrac{5xy\sqrt{yz}}{7z^3}$

   (e) $\dfrac{4\sqrt{3}}{9}$ (f) $\dfrac{\sqrt{85}}{17}$ (g) $\dfrac{\sqrt{3xy}}{3y}$ (h) $\dfrac{\sqrt{34xy}}{10y^2}$

   (i) $\dfrac{\sqrt{21}}{3}$ (j) $\dfrac{\sqrt{10}}{5}$ (k) 3   (l) $\dfrac{\sqrt{ab}}{b}$

7. for #1 (b) $2\sqrt{38}$ (c) $\dfrac{\sqrt{989}}{5}$ (d) $x^2 y^3 \sqrt{55}$

         (e) $\dfrac{\sqrt{21}}{7}$ (g) $9\sqrt{\frac{2}{7}}$ (h) $\dfrac{xy^4}{21z^4}\sqrt{105x}$

   for #2 (a) $\dfrac{\sqrt{143}}{13}$ (b) $\dfrac{\sqrt{21}}{3}$ (c) $\frac{3}{5}\sqrt{29}$ (d) $2x^3 y^8 \sqrt{3}$

         (e) $\dfrac{\sqrt{77}}{11}$ (f) $\dfrac{\sqrt{30}}{2}$ (g) $\frac{2}{5}\sqrt{3}$ (h) $\dfrac{x^3 y^2 \sqrt{15}}{5z}$

9. (a) $\dfrac{7 - 2\sqrt{10}}{3}$ (b) $\dfrac{29 + 7\sqrt{21}}{4}$

Section 3

1. (a) $\sqrt[3]{92}$ (b) $\sqrt[4]{48}$ (c) $\sqrt[5]{480}$ (d) $\sqrt[3]{.008}$
   (e) $\sqrt[3]{4x^3 y^{11}}$ (f) $\sqrt[5]{288x^7 y^7 z^{12}}$ (g) $\sqrt[3]{\frac{10}{27}}$ (h) $\sqrt[5]{\frac{45}{64}}$
   (i) $\sqrt[3]{162x^4 y^7}$ (j) $\sqrt[7]{128x^8 y^{11}}$

3. (a) $3\sqrt[3]{2}$ (b) $2xy^2\sqrt[4]{3x^2y^2}$ (c) $.1\sqrt[5]{.1}$

(d) $-2\sqrt[3]{2}$ (e) $-\sqrt[3]{17}$ (f) $\dfrac{\sqrt[3]{80}}{4}$

(g) $\dfrac{\sqrt[4]{45}}{3}$ (h) $\dfrac{\sqrt[3]{5x^2}}{x}$ (i) $2x^5y^6z^7\sqrt{2x^4y^5z^4}$

(j) $-3x^2y^3\sqrt[3]{2xy^2z}$ (k) $\dfrac{\sqrt[3]{12}}{2}$ (l) $\dfrac{\sqrt[5]{16}}{2}$

5. for #1 (b) $2\sqrt[4]{3}$ (c) $2\sqrt[5]{15}$ (d) $.2$

(e) $xy^3\sqrt[3]{4y}$ (f) $2xyz^2\sqrt[5]{9x^2y^2z^2}$ (g) $\dfrac{\sqrt[3]{10}}{3}$

(h) $\dfrac{\sqrt[5]{45}}{2\sqrt[5]{2}}$ (i) $3xy^2\sqrt[3]{6xy}$ (j) $2xy\sqrt[7]{xy^4}$

for #2 (a) $\frac{1}{4}\sqrt[4]{4}$ (c) $\sqrt[5]{5}/\sqrt[5]{12}$ (d) $.3\sqrt[3]{2}$

(e) $ab^3\sqrt[3]{4b^2}$ (f) $2abc^2\sqrt[5]{18a^3bc^3}$ (g) $\sqrt[3]{14/3}$

(h) $\sqrt[5]{63}/2\sqrt[5]{5}$ (i) $3ab^3\sqrt[3]{10a}$ (j) $ay\sqrt[7]{320ay^2}$

7. (a) $\dfrac{\sqrt[3]{49}+\sqrt[3]{14}+\sqrt[3]{4}}{5}$ (b) $2(\sqrt[3]{25}-\sqrt[3]{10}+\sqrt[3]{4})$

(c) $\dfrac{\sqrt[3]{a^2}+\sqrt[3]{ab}+\sqrt[3]{b^2}}{a-b}$ (d) $\dfrac{\sqrt[3]{a^2}-\sqrt[3]{ab}+\sqrt[3]{b^2}}{a+b}$

Section 4

1. (a) 6.557   (b) 9.534   (c) 3.849   (d) 3.606   (e) 3.391
(f) 8.775
3. (a) 68.56   (b) 0.6708   (c) 0.8944   (d) 0.2410   (e) 97.98
(f) 0.3583   (g) 20.78   (h) 13.23   (i) 4.703   (j) 6.672
(k) 20.12   (l) 7.211
5. (a) 1.291   (b) 1.783   (c) 1.363   (d) 4.301   (e) 1.206
(f) 1.205   (g) 1.327   (h) 3.391   (i) 1.363   (j) 0.9608
(k) 0.9565   (l) 1.367
7. (a) 2.447   (b) 4.266   (c) 1.828   (d) 1.081   (e) -5.663
(f) 6.599
9. 3.48 in.   11. 5.143 cm

Section 5

1. (a) 2   (b) $\frac{1}{2}$ (c) 8   (d) $\frac{1}{8}$ (e) 4   (f) $\frac{1}{4}$ (g) 32

(e) $\dfrac{\sqrt{2}}{4}$

3. (a) 2   (b) $8^{1/4}$ (c) 0.1   (d) $\frac{3}{2}$ (e) $5^{1/3}/2$
(f) $8/\sqrt{27}=8\sqrt{27}/27$

5. (a) $2xy^2$ (b) $2xy^2\sqrt{5xy}$ (c) $3x^3y^4\sqrt{2x}$
   (d) $3x^2y^3\sqrt[3]{xy^2}$ (e) $8x^3y^6$ (f) $4x^4y^6$
   (g) $.04x\sqrt[3]{x}$ (h) $3.2(10)^6x^{10}$ (i) $16x^{8/9}y^{16/15}$
   (j) $5x^{2/3}y^{5/4}$

CHAPTER 5

Section 1

1. (a) $1, 6$ (b) $9, 4$ (c) $7, 5$ (d) $7, -1$ (e) $-5, 3$
   (f) $-1, \frac{3}{2}$ (g) $-\frac{7}{2}, \frac{1}{3}$ (g) ± 2

3. (a) $\frac{3}{2}, -\frac{1}{4}$ (b) $\dfrac{-8 \pm \sqrt{118}}{18}$ (c) $0, -\frac{1}{2}, 7$ (d) $-3, 7$

   (e) $0, 3, -\frac{5}{2}$ (f) $0, \frac{3}{2}, -\frac{5}{6}$ (g) $x = \pm 2$ (h) $x = \pm 3$
5. (a) ± 5 (b) ± 7 (c) $\pm \frac{1}{3}$ (d) $\pm \sqrt{7}$
   (e) $\pm \sqrt{7/3}$ (f) $\pm \sqrt{k}$ (g) $4, -6$ (h) $9, -5$

   (i) $-\frac{8}{5}, -\frac{10}{3}$ (j) $4 \pm \sqrt{7}$ (k) $\dfrac{-15 \pm \sqrt{7}}{3}$ (l) $p = \sqrt{k}$

7. (a) $-7, 3$ (b) $-1, 7$ (c) $-3 \pm \sqrt{5}$ (d) $-h \pm a$
   (e) $k \pm b$ (f) $-r \pm \sqrt{c}$
9. (a) $\pm 3, \pm 2$ (b) $\pm 1, \pm 4$ (c) $\pm 2, \pm 5$ (d) $\pm a, \pm b$

Section 2

1. $-2 \pm 2\sqrt{3}$ 3. $-1 \pm 2\sqrt{5}$ 5. $\dfrac{3 \pm \sqrt{17}}{2}$

7. $-4 \pm 3\sqrt{2}$ 9. $\dfrac{-3 \pm \sqrt{19}}{2}$ 11. no solution

13. $-4 \pm 2\sqrt{5}$ 15. $1, \frac{2}{11}$ 17. $-1, \frac{1}{4}$

19. $\dfrac{-1 \pm \sqrt{193}}{8}$ 21. $\dfrac{4 \pm 2\sqrt{31}}{3}$ 23. $\dfrac{-13 \pm 3\sqrt{17}}{4}$

25. no solution 27. $\dfrac{-3 \pm \sqrt{177}}{12}$ 29. $\dfrac{-20 \pm \sqrt{406}}{3}$

Section 3

1. (a) $-2 \pm 3\sqrt{2}$ (b) $\dfrac{7 \pm \sqrt{61}}{2}$ (c) $-1, \frac{2}{3}$

   (d) $\dfrac{-6 \pm \sqrt{46}}{2}$ (e) $-\frac{5}{2}, 1$ (f) $\dfrac{-1 \pm \sqrt{22}}{3}$

   (g) $\dfrac{-1 \pm \sqrt{33}}{8}$ (h) $-1, \frac{1}{3}$

3. (a) $\dfrac{-3 \pm \sqrt{41}}{4}$ (b) $\dfrac{-5 \pm \sqrt{33}}{2}$ (c) $\dfrac{-1 \pm \sqrt{253}}{14}$

(d) $\dfrac{-1 \pm \sqrt{11}}{12}$ (e) $\dfrac{-7 \pm \sqrt{29}}{10}$ (f) $\dfrac{-\sqrt{35} \pm \sqrt{231}}{14}$

(g) $\dfrac{-17 \pm \sqrt{2081}}{28}$ (h) $\dfrac{15 \pm \sqrt{33}}{32}$ (i) no solution

(j) no solution

7. (a) $\left(x + \dfrac{5 - \sqrt{65}}{2}\right)\left(x + \dfrac{5 + \sqrt{65}}{2}\right)$

(b) $3\left(x + \dfrac{1 - \sqrt{22}}{3}\right)\left(x + \dfrac{1 + \sqrt{22}}{3}\right)$

(c) $5\left(x - \dfrac{3 + \sqrt{29}}{10}\right)\left(x - \dfrac{3 - \sqrt{29}}{10}\right)$

(d) $16\left(x - \dfrac{17 + \sqrt{161}}{32}\right)\left(x - \dfrac{17 - \sqrt{161}}{32}\right)$

(e) $7\left(x + \dfrac{3 - \sqrt{541}}{14}\right)\left(x + \dfrac{3 + \sqrt{541}}{14}\right)$

(f) $4x\left(x + \dfrac{7 - \sqrt{33}}{8}\right)\left(x + \dfrac{3 + \sqrt{33}}{8}\right)$

(g) $5\left(x - \dfrac{3 + \sqrt{-31}}{10}\right)\left(x - \dfrac{3 - \sqrt{-31}}{10}\right)$

(h) $0.4\left(x + \dfrac{25 - \sqrt{673}}{4}\right)\left(x + \dfrac{25 + \sqrt{673}}{4}\right)$

9. (a) $-\frac{5}{2}$ (b) $-\frac{9}{2}$ (c) $-\frac{5}{8}$ (d) $-\frac{5}{6}$ (e) $-\frac{7}{6}$ (f) $-\frac{3}{10}$
   (g) $-\frac{5}{8}$ (h) $-\frac{1}{6}$

11. (b) $-\frac{3}{4}$ (c) $\frac{13}{16}$ (d) $-\frac{1}{24}$ (f) $\dfrac{-\sqrt{6}}{6}$ (g) $\frac{7}{18}$

(h) $\frac{5}{46}$ (i) $-\frac{11}{24}$ (j) $\frac{13}{32}$

Section 4

1. (a) $-4 < x < 5$ (b) $x < -2$ or $x > 7$
   (c) $x < -\frac{3}{2}$ or $x > 4$ (d) $x < -3$ or $0 < x < 2$
   (e) $x < -4$ or $x > 1$ (f) $x < 2$ or $x > 5$

(g) $\dfrac{-9 - \sqrt{161}}{2} < x < \dfrac{-9 + \sqrt{161}}{2}$ (h) $x < -7$ or $x > 1$

(i) $-\frac{5}{2} < x < 7$ (j) $-10 < x < 0$ or $x > 2$
(k) $x \neq 2$ (l) no solution

3. (a) $-2 < x < 0$ or $x > 3$ (b) $x < -1$ or $1 < x < 2$

(c) $x < -7$, $-5 < x < 4$, or $x > 9$ (d) $-6 < x < -3$ or $7 < x < 11$

(e) $x < -9$, $-6 < x < -2$, $0 < x < 4$ or $x > 5$

(f) $x < -5$ or $x > 7$ (g) $x < -3$ or $x > 4$

(h) $-7 < x < 1$ or $1 < x < 8$

(i) $x < -5$, $-5 < x < -2$, $3 < x < 7$, or $x > 7$

(j) $x < -2$, $-2 < x < 3$, $3 < x < 7$, or $x > 7$

5. (a) $x < p$ or $x > q$ (b) $p < x < q$

7. (a) $\{x \mid x < q \text{ or } x > r\} - p$ (b) $\{x \mid q < x < r\} - p$

Section 5

1. (a) $x < -7$ or $x > 7$ (b) $-8 < x < 8$ (c) $-10 < x < 4$

(d) $x > 13$ or $x < -3$ (e) $x > 3$ or $x < -4$ (f) $-\frac{1}{3} < x < 3$

(g) $-\frac{14}{5} < x < 0$ (h) $-\frac{13}{6} < x < \frac{1}{2}$ (i) $x \ne -\frac{2}{3}$

(j) no solution

3. (a) $x < -\sqrt{13}$ or $x > \sqrt{13}$ (b) $-\sqrt{17} < x < \sqrt{17}$

(c) $-2 - \sqrt{13} < x < -2 + \sqrt{13}$ (d) $x < 7 - \sqrt{17}$ or $x > 7 + \sqrt{17}$

(e) $\dfrac{3 - \sqrt{13}}{2} < x < \dfrac{3 + \sqrt{13}}{2}$

(f) $x < \dfrac{-2 - \sqrt{17}}{3}$ or $x > \dfrac{-2 + \sqrt{17}}{3}$

(g) $x < \dfrac{-6 - \sqrt{13}}{5}$ or $x > \dfrac{-6 + \sqrt{13}}{5}$

(h) $\dfrac{3 - \sqrt{17}}{7} < x < \dfrac{3 + \sqrt{17}}{7}$ (i) all x (j) no solution

5. (a) $x < -3 - \sqrt{17}$ or $x > -3 + \sqrt{17}$ (b) $2 - \sqrt{14} < x < 2 + \sqrt{14}$

(c) $x < 5 - \sqrt{37}$ or $x > 5 + \sqrt{37}$ (d) $-4 - \sqrt{13} < x < -4 + \sqrt{13}$

(e) $\dfrac{-2 - 3\sqrt{2}}{2} < x < \dfrac{-2 + 3\sqrt{2}}{2}$

(f) $x < \dfrac{-9 - \sqrt{177}}{6}$ or $x > \dfrac{-9 + \sqrt{177}}{6}$

(g) $x < -1$ or $x > \frac{3}{5}$ (h) $\dfrac{-1 - \sqrt{31}}{6} < x < \dfrac{-1 + \sqrt{31}}{6}$

(i) all x (j) no solution

7. $x < \dfrac{-B - \sqrt{B^2 - 4AC}}{2A}$ or $x > \dfrac{-B + \sqrt{B^2 - 4AC}}{2A}$ if $B^2 - 4AC > 0$

9. $Ap^2 + Bp + C \le 0$

Section 6

1. $x = 2 + 2\sqrt{6}$, $y = -2 + 2\sqrt{6}$ or $x = 2 - 2\sqrt{6}$, $y = -2 - 2\sqrt{6}$
3. 15.81   5. $2\sqrt[3]{2}$ in. New volume 8 times the original volume.
7. area of square 625 sq. ft

area of circle $\dfrac{2500}{\pi} = 795.8$ sq. ft or if $\dfrac{2500}{\pi} < 625$, then $\dfrac{2500}{625} = 4 < \pi$, which is
a contradiction

9. (a) $R = 0.0042$ (b) $C = \$60.48$
11. (a) $C = 500$ (b) $I = 5$ units   (c) $2\sqrt{5}$ cm
13. (a) \$600   (b) $S > 140$ units   (c) $S > 240$ units

15. $(45 - 5\sqrt{41})$ ft   17. (a) $\dfrac{5 - \sqrt{15}}{2}$ sec   (b) no, max height is 100 ft

CHAPTER 6

Section 2

21. (i) In each case the answer is $-\frac{3}{2}$.   (ii) $y = -\frac{3}{2}x + \frac{5}{2}$ with slope $-\frac{3}{2}$.

Section 3

1. $y = 5x + 3$ 3. $y - 2 = 7(x - 5)$ 5. $y - 9 = -8(x - 2)$
7. $y - 4 = 4(x + 3)$ 9. $y - 13 = \frac{2}{3}(x - 10)$ 11. $y - \frac{2}{7} = 15(x - \frac{1}{3})$
13. $y - 1 = -\frac{2}{5}(x - 7)$ or $y - 3 = -\frac{2}{5}(x - 2)$
15. $y = 4x + 4$ or $y - 8 = 4(x - 1)$
17. $y + 2 = 3(x - 5)$ or $y - 7 = 3(x - 8)$
19. $y - 19 = -\frac{2}{7}(x + 4)$ or $y - 13 = -\frac{2}{7}(x - 17)$
21. $y + \frac{1}{3} = -\frac{5}{3}(x - \frac{1}{2})$ or $y - \frac{1}{6} = -\frac{5}{3}(x - \frac{1}{5})$
23. $y = 4.1 = \frac{60}{37}(x - .13)$ or $y - 5.3 = \frac{60}{37}(x - .87)$
25. $y - .4 = -\frac{54}{13}(x + .6)$ or $y + 5 = -\frac{54}{13}(x - .7)$
27. $x = 4$ 29. $y - 5 = 5(x - 1)$ 31. $y - 3 = -\frac{3}{2}(x - 2)$
33. $y = 3$ 35. $x = 7$

Section 4

1. (a) $y = -6x + 19$ (b) -29
3. (a) $y = -1200x + 3200$ (b) 1400 gal   (c) 800 gal
5. (a) $y = -\frac{25}{2}(x - 5)$ (b) If $x = 2\text{¢}$, $y = 37.5\%$
                                       $x = 3\text{¢}$, $y = 25\%$
                                       $x = 4\text{¢}$, $y = 12.5\%$
7. (a) $y = \frac{1}{8}x - 29{,}500$ (b) $x = 236{,}000$
    (c) loss of \$10,750 or $y = -10{,}750$ dollars

Section 5

1. (a) $\sqrt{26}$ (b) 5   (c) $\sqrt{73}$ (d) $\sqrt{73}$ (e) $\sqrt{2}$
    (f) $2\sqrt{10}$ (g) $2\sqrt{a^2 + b^2}$ (h) $(\sqrt{2} - 1)\sqrt{a^2 + b^2}$
5. the single point $(0, 0)$

Section 6

1. $(\frac{47}{26}, -\frac{9}{26})$

3. $\left(\dfrac{4 + 2\sqrt{19}}{5}, \dfrac{2 - 4\sqrt{19}}{5}\right), \left(\dfrac{4 - 2\sqrt{19}}{5}, \dfrac{2 + 4\sqrt{19}}{5}\right)$

5. $\left(\dfrac{24 + 9\sqrt{23}}{13}, \dfrac{10 - 6\sqrt{23}}{13}\right), \left(\dfrac{24 - 9\sqrt{23}}{13}, \dfrac{10 + 6\sqrt{23}}{13}\right)$

7. $\left(\dfrac{53 + 25\sqrt{33}}{34}, \dfrac{77 - 15\sqrt{33}}{34}\right), \left(\dfrac{53 - 25\sqrt{33}}{34}, \dfrac{77 + 15\sqrt{33}}{34}\right)$

9. $\left(\dfrac{-3 + \sqrt{17}}{2}, \dfrac{13 - 3\sqrt{17}}{2}\right), \left(\dfrac{-3 - \sqrt{17}}{2}, \dfrac{13 + 3\sqrt{17}}{2}\right)$

11. $(\frac{1}{2}, \frac{25}{4})$

Section 7

1. (a) $D_f = \{x: -5 < x < 5\}$ (b) $D_g = \{x: x \neq 1\}$
$R_f = \{y: 0 \leq y \leq 5\}$ $R_g = R$
(c) $D_h = R$ (d) $D_k = R$
$R_h = \{y: 0 < y \leq 9\}$ $R_n = R$
(e) $D_p = \{x: x \neq -1, x \neq 4\}$
$R_p = R$
(f) $D_q = \{x: x < -4, -4 < x \leq -\sqrt{5}, \sqrt{5} \leq x < 4, \text{ or } x > 4\}$
$R_g = R$

11. (a) $f(x - a)$ is the graph of $f(x)$ shifted along the x axis a units
(b) $f(x) + b$ is the graph of $f(x)$ shifted vertically

Section 8

1. $C(x) = 7.5x + 300$ where x is the number of units produced
$C(275) = \$2,362.50$.
3. $E(x) = 12x + 100$ where x is the unemployment rate in percent
$E(6.5) = 178$
5. $P(0) < 0$ 9. $G(x) = 0.952x + 1.982$

Section 9

1. (a) $x^3 + x^2 + 1$ (b) $x^5 + x^3$ (c) $\dfrac{x^3}{\sqrt{x - 2}}$
$D_{f+g} = R$ $D_{fg} = R$
$D_{g/h} = \{x \mid x > 2\}$

(d) $(x + 1)\sqrt{x - 2}$ (e) $\dfrac{x^2 + 1}{x + 1}$ (f) $x^3 - x^2 - 1$
$D_{hk} = \{x: x > 2\}$ $D_{fk} = R - \{-1\}$ $D_{gf} = R$

(g) $x^4 + x^3$

     $D_{g/f} = R$

(h) $\dfrac{1}{x^3 + x^2 + x + 1}$

     $D_{k/f} = R - \{-1\}$

3. (a) $x^6 + 1$

     $D_{f \circ g} = R$

(b) $x^6 + 3x^4 + 3x^2 + 1$

     $D_{g \circ f} = R$

(c) $x - 1$

     $D_{f \circ h} = \{x : x \geq 2\}$

(d) $\sqrt{\dfrac{-(2x + 1)}{x + 1}}$

     $D_{h \circ k} = \{x : -1 < x < -\tfrac{1}{2}\}$

(e) $\dfrac{x^2 + 2x + 2}{x^2 + 2x + 1}$

     $D_{f \circ k} = \{x : x \neq -1\}$

(f) $(x - 2)\sqrt{x - 2}$

     $D_{g \circ h} = \{x : x \geq 2\}$

(g) $\sqrt{\dfrac{-x(x + 2)}{(x + 1)^2}}$

     $D_{(h \circ f) \circ k} = \{x : -2 \leq x < -1 \text{ or } -1 < x \leq 0\}$

(h) $\dfrac{x^3 + 3x^2 + 3x + 1}{x^3 + 3x^2 + 3x + 2}$

     $D_{h \circ (g \circ h)} = \{x : x < -2, \ -2 < x < -1, \ x > -1\}$

Section 11

13. (a) $\dfrac{-B + \sqrt{B^2 - 4AC}}{2A}, \ \dfrac{-B - \sqrt{B^2 - 4AC}}{2A}$

(b) $B^2 - 4AC < 0$

(c) If $A > 0$ it is above the x-axis.

     If $A < 0$ it is below the x-axis.

APPENDIX

1. (a) -151 (b) -18 (c) -88 (d) 4.5 (e) $-\frac{13}{12}$

     (f) $\frac{688}{7}$ (g) -118 (h) -15.8 (i) -80.2 (j) $\frac{101}{150}$

3. (a) -275.2 (b) 251.68 (c) -5460 (d) -82.49

     (e) 0.27888 (f) $-\frac{987}{320}$ (g) $-\frac{138}{7}$ (h) -37

     (i) $\frac{125}{16}$ (j) -35 (k) $\frac{99}{8}$ (l) $\frac{3}{8}$

Index

Absolute value, 27
Addition and subtraction of algebraic
   fractions, 108
Algebraic expression, 79
Applications of functions, 238
Arithmetic
   of functions, 248
   of polynomials, 80
   of signed numbers, 287
Asymptote, 279
Axiom of extension, 8
Axioms for the real numbers, 13

Binomials, 84

Cancelling, 101
Center
   of an ellipse, 261, 262
   of a hyperbola, 279, 283
Checking by substitution, 37
Classification of numbers, 30
Coefficients of a polynomial, 84
Collecting like terms, 84

Completing the square, 159
Composite of functions, 254
Conditional equation, 34
Constants, 3
Coordinates, 193
Coordinate system, 193
Cube root, 135

Decimal approximations of numbers
   expressed using radicals, 139
Degree of a polynomial, 113
Difference of sets, 7
Discriminant of a quadratic equation,
   171
Disjoint sets, 7
Distance formula, 217
Division algorithm for polynomials, 114
Division by zero, 20
Domain of a function, 230

Element of a set, 4
Ellipse, 259
Empty set, 6

Equality, 8
Equation of the circle of radius r with
    center (h, k), 220
Equations of horizontal and vertical
    lines, 200
Equations of lines, 206
Exponents, 80

Factoring polynomials, 91
Factor theorem, 115
Function, 227, 231
Fundamental laws governing exponents
    and radicals, 151
Fundamental theorem of arithmetic, 31

Graph
    concept of, 193
    of an equation, 61
    of a function, 232
    of a system of equations, 222
Greater than, 15

Hyperbola, 274

Identity, 34
Index of a radical, 135
Integers, 30
Integral exponents, 117
Intersection of sets, 6

Language of sets, 4
Law
    of exponents, 83, 119
    of signs, 18
Least common denominator, 108
Least complicated common
    denominator, 108
Least upper bound property, 17
Less than, 15
Like terms, 84
Linear equations in one variable, 33
Linear variation, 212
Lines, 199
Literal equations
    and formulas, 43
Lowest terms, 31

Mapping, 227

Monomials, 84
Multiplication and division of
    algebraic fractions, 104

Natural numbers, 30
Negative number, 15
Null set, 6
Numerals, 34

Ordered pairs of numbers, 194
Order properties of R, 23

Parabola, 265
Perfect square, 140
Pi (π), 3
Piecewise linear function, 243
Plot and sketch technique for
    graphing, 223
Point–slope form of the equation of a
    line, 209
Polynomial, 84
Polynomial equation in one variable,
    153
Polynomial inequalities, 172
Positive number, 31
Prime number, 31
Principle nth root, 135
Principle square root, 127
Problem solving, 69
Product of two binomials, 89
Proper subset, 5

Quadratic equation, 159
Quadratic formula, 168

Radical 128, 135
Radical sign, 128, 135
Radicand, 128, 135
Range, 230
Rational exponents, 145
Rational numbers, 30
Real numbers, 13
Remainder theorem, 115
Rise, 208
Run, 208

Scientific notation, 124
Set, 4
Set builder notation, 5

Simplifying algebraic fractions, 99
Simplifying radicals, 129, 136
Slope of a line, 202
Slope–intercept form of the equation of
     a line, 202
Solution
   of an equation, 34
   of linear inequalities, 58
   of polynomial equations by
     factoring, 153
   of quadratic inequalities, 178
   of systems of equations, 47
Square root, 127
Straight line, 199
Subset, 5
Substitution principle, 8
Subtraction, 19
Systems, of equations
   in three or more variables, 54
   in two variables, 46

Terms of a polynomial, 84
Trinomial, 84
Two-point form of the equation of a
     line, 209

Union of sets, 6
Universal set, 9
Universe of discourse, 9

Variable, 3, 34
Venn diagram, 9
Vertex
   of an ellipse, 261, 262
   of a hyperbola, 279, 283
   of a parabola, 267, 270, 272

y-intercept, 202

Index of Symbols

π, 3	$\geq$, 15	$f(x)$, 113
$\in$, 4	$<$, 15	a^0, 118
$\subseteq$, 5	$\leq$, 15	a^{-n}, 119
$\nsubseteq$, 5	a/b, 16	$\sqrt{}$, 127
$\subsetneqq$, 5	$a - b$, 19	$\sqrt[n]{}$, 135
$\varnothing$, 6	N, 30	$a^{1/n}$, 147
$\cup$, 6	Z, 30	$a^{p/q}$, 149
$\cap$, 6	Q, 30	$f(a)$, 229
$A - B$, 7	R, 30	$f \circ g$, 254
$>$, 15	a^n, 81	

77 78 79 80 9 8 7 6 5 4 3 2 1